PENGUIN BOOKS

THE PENGUIN DICTIONARY OF
CURIOUS AND INTERESTING GEOMETRY

David Wells was born in 1940. He had the rare distinction of being a Cambridge scholar in mathematics and failing his degree. He subsequently trained as a teacher and, after working on computers and teaching machines, taught mathematics and science in a primary school and mathematics in secondary schools. He is still involved with education through writing and research.

While at university he became British under-21 chess champion, and in the mid-seventies was a game inventor, devising 'Guerrilla' and 'Checkpoint Danger', a puzzle composer, and the puzzle editor of *Games & Puzzles* magazine. From 1981 to 1983 he published *The Problem Solver*, a magazine of mathematical problems for secondary pupils. He has published several books of problems and popular mathematics, including *Can You Solve These?* and *Hidden Connections, Double Meanings*, as well as *Russia and England, and the Transformation of European Culture*. He is also the author of *The Penguin Book of Curious and Interesting Puzzles*, *The Penguin Dictionary of Curious and Interesting Numbers*, *The Penguin Dictionary of Curious and Interesting Geometry*, *The Penguin Book of Curious and Interesting Mathematics* and *You Are a Mathematician*.

The Penguin Dictionary of Curious and Interesting Geometry

DAVID WELLS

illustrated by
JOHN SHARP

PENGUIN BOOKS

PENGUIN BOOKS

Published by the Penguin Group
Penguin Books Ltd, 27 Wrights Lane, London W8 5TZ, England
Penguin Putnam Inc., 375 Hudson Street, New York, New York 10014, USA
Penguin Books Australia Ltd, Ringwood, Victoria, Australia
Penguin Books Canada Ltd, 10 Alcorn Avenue, Toronto, Ontario, Canada M4V 3B2
Penguin Books (NZ) Ltd, Private Bag 102902, NSMC, Auckland, New Zealand

Penguin Books Ltd, Registered Offices: Harmondsworth, Middlesex, England

First published 1991
7 9 10 8

Printed in England by Clays Ltd, St Ives plc
Typesetting in Linotron 10/13 Sabon

Contents

Introduction

Circles, rectangles, triangles and spirals are found in prehistoric art and in the art and decorations of primitive man. Even before human beings entered the scene they were found in nature, as were innumerable crystals, so perfectly and mysteriously geometrical that it was believed until recently that they grew in the earth according to some vital principle.

Egyptian architecture displays many geometrical forms and features, and an early style of Greek art is called geometric from the patterns displayed. As soon as the Greeks started to look at geometrical figures for their own sake, a new wealth of properties was revealed. The Pythagorean triangle, on the other hand, is far older than Pythagoras. It may be as old as the stone age.

When Menæchmus sliced a cone, figures were revealed that two thousand years later proved to be one key to the motion of the planets. When Archimedes found volumes by summing many parallel slices, he was anticipating the integral calculus.

Many of the most important advances in the history of mathematics have been achieved by leaps of geometrical insight not excluding the ordinary and familiar. Ironically, topologists were the first to look with mathematical eyes on the humble knot, which is as old as history itself.

Most recently, the study of fractals and chaos has revealed images of unexpected beauty, depth and mystery, as well as exhibiting the continuing power of geometrical styles of thinking in the physical sciences.

This is a companion to the *Penguin Dictionary of Curious and Interesting Numbers*, with a difference, however. The variety of geometrical images is so great that no one book could contain more than a sampling. Entire books have been written about tessellations alone, or topological curiosities, or geometrical extremal properties, beside the wealth of classical geometry. This is my selection from that cornucopia.

Many of the entries are identified by the name of the discoverer (or the name which popular history has attached to them – not always the same person!) All the names mentioned will be found in the index, with dates and domiciles where appropriate. A small number of recent sources are credited in references to particular journals or books.

Further information, and many of the entries that I would like to have included but could not, will be found in the books listed in the bibliography. May I add, however, that I do hope that readers will be at least as keen to take up pen and paper and investigate ideas that intrigue them for themselves, as they will be to search further sources. Geometry, like number theory, like all of mathematics, should not be a spectator sport!

I am very grateful to several copyright holders for permission to use diagrams from their books or journals. These are recorded below.

I should like to thank David Singmaster once again for the use of his extensive library; Peter Mayer for his helpful suggestions; John O'Driscoll for the hand drawn figures; and Ravi Mirchandani of Penguin Books for his enthusiasm for and patient oversight of this dictionary.

Finally, I would like to thank John Sharp for producing the illustrations by computer, in many cases improving on their traditional presentation and producing some which have never been seen before.

ACKNOWLEDGEMENTS

The author and publishers would like to thank the following for permission to reproduce illustrations: R. DIXON, *Mathographics*, pp 165–166, Basil Blackwell, Oxford, 1987, in the entries on the Fermat spiral and inversion; P. DO CARMO, *Differential Geometry of Curves and Surfaces*, pp 223–224, Prentice Hall, Englewood Cliffs, New Jersey, 1976, in the entry on helicoids; MARTIN GARDNER, *The Sixth Book of Mathematical Games from Scientific American*, W. H. Freeman, San Francisco, 1971, in the entry on the billiard ball path in a cube; D. HILBERT and S. COHN-VOSSEN, *Geometry and the Imagination*, p 23, Chelsea Publishing Company, New York, 1952, in the entry on orthogonal surfaces; DAVID WELLS, *Hidden Connections, Double Meanings*, p 31, Cambridge University Press, 1988, in the entry on Haüy's construction of polyhedra; The Mathematical Association of America, *Mathematics Magazine*, vol 52 (1), January 1979, p 13, for the Chinese illustration of Pythagoras' theorem.

A Chronological List Of Mathematicians

This list includes all the important mathematicians named in this dictionary, other than those still living, plus several scientists and others, such as Leonardo and Galileo. It is surprising how many well-known mathematicians are known to non-mathematicians as physicists, engineers, and so on!

Thales of Miletus	*c*.625–*c*.547 BC	Greek
Pythagoras	*c*.580–*c*.480 BC	Greek
Hippocrates of Chios	fl. *c*.440 BC	Greek
Plato	*c*.427–347 BC	Greek
Aristotle	384–322 BC	Greek
Euclid	fl. *c*.295 BC	Greek
Philo	fl. *c*.250 BC	Greek
Nicomedes	fl. *c*.240 BC	Greek
Perseus	fl. 3rd cent BC	Greek
Archimedes	*c*.287 BC–212 BC	Greek
Diocles	*c*.180 BC	Greek
Apollonius of Perga	*c*.225 BC–*c*.175 BC	Greek
Heron of Alexandria	fl. *c*. 62 AD	Greek
Menelaus of Alexandria	fl. *c*. 100	Greek
Ptolemy	*c*.85–*c*.165	Greek
Pappus of Alexandria	fl. 300–350	Greek
Abu'l Wefa	940–998	Persian
Regiomontanus, Johannes	1436–1476	German
Pacioli, Luca	*c*.1445–1517	Italian

Leonardo da Vinci	1452–1519	Italian
Dürer, Albrecht	1471–1528	German
Galileo (Galileo Galilei)	1564–1642	Italian
Kepler, Johann	1571–1630	German
Mersenne, Marin	1588–1648	French
Pascal, Étienne	1588–1651	French
Desargues, Girard	1591–1661	French
Descartes, René du Perron	1596–1650	French
Fermat, Pierre de	1601–1665	French
Roberval, Gilles Personne de	1602–1675	French
Torricelli, Evangelista	1608–1647	Italian
Schooten, Frans van	1615–1660	Dutch
Pascal, Blaise	1623–1662	French
Cassini, Giovanni Domenico	1625–1712	Italian
Huygens, Christiaan	1629–1695	Dutch
Wren, Christopher	1632–1723	English
Mohr, Georg	1640–1697	Danish
Newton, Isaac	1642–1727	English
Leibniz, Gottfried Wilhelm	1646–1716	German
Ceva, Giovanni	1647/8–1734	Italian
Tschirnhausen, Ehrenfried Walther von	1651–1708	German
Bernoulli, Jakob	1654–1705	Swiss
Simson, Robert	1687–1768	Scottish
Bernoulli, Daniel	1700–1782	Swiss
Euler, Leonhard	1707–1783	Swiss
Malfatti, Gian Francesco	1731–1807	Italian
Lagrange, Joseph Louis	1736–1813	Italian
Watt, James	1736–1819	Scottish
Haüy, René-Just	1743–1822	French
Monge, Gaspard	1746–1818	French
Mascheroni, Lorenzo	1750–1800	Italian

Carnot,		
Lazare Nicolas Marguerite	1753–1823	French
Gergonne, Joseph Diez	1771–1859	French
Bowditch, Nathaniel	1773–1838	American
Gauss, Carl Friedrich	1777–1855	German
Poinsot, Louis	1777–1859	French
Crelle, August Leopold	1780–1855	German
Brianchon, Charles Julien	1783–1864	French
Poncelet, Jean Victor	1788–1867	French
Cauchy, Augustin Louis	1789–1857	French
Möbius, August Ferdinand	1790–1868	German
Lobachevsky, Nikolai Ivanovich	1792–1856	Russian
Dandelin, Germinal Pierre	1794–1847	Belgian
Steiner, Jakob	1796–1863	Swiss
Feuerbach, Karl Wilhelm	1800–1834	German
Plücker, Julius	1801–1868	German
Plateau,		
Joseph Antoine Ferdinand	1801–1883	Belgian
Bolyai, János	1802–1860	Hungarian
Verhulst, Pierre-François	1804–1849	Belgian
Jacobi, Carl Gustav Jacob	1804–1851	German
Kirkman, Thomas Penyngton	1806–1895	English
Schläfli, Ludwig	1814–1895	Swiss
Salmon, George	1819–1904	Irish
Cayley, Arthur	1821–1895	English
Lissajous, Jules Antoine	1822–1880	French
Cremona, Antonio		
Luigi Gaudenzio Giuseppe	1830–1903	Italian
Beltrami, Eugenio	1835–1899	Italian
Reye, Theodor	1838–1919	German
Lemoine,		
Émile Michel Hyacinthe	1840–1912	French
Neuberg, Joseph	1840–1926	Belgian

Schwarz, Hermann Amandus	1843–1921	German
Clifford, William Kingdom	1845–1879	English
Brocard, Pierre René Jean-Baptiste Henri	1845–1922	French
Dudeney, Henry Ernest	1847–1930	English
Klein, Christian Felix	1849–1925	German
Poincaré, Jules Henri	1854–1912	French
Föppl, August	1854–1924	German
Morley, Frank	1860–1937	American
Hilbert, David	1862–1943	German
Kürschák, József	1864–1933	Hungarian
Koch, Helge von	1870–1924	Swedish
Fano, Gino	1871–1952	Italian
Lebesgue, Henri Léon	1875–1941	French
Soddy, Frederick	1877–1956	English
Fatou, Pierre Joseph Louis	1878–1929	French
Sommerville, Duncan Mclaren Young	1879–1934	Scottish
Sierpinski, Waclaw	1882–1969	Polish
Thébault, Victor	1882–1960	French
Blashke, Wilhelm Johann Eugen	1885–1962	Austrian
Julia, Gaston	1893–1978	French

Bibliography

This list is limited to books entirely devoted to geometry. Of the many books which include some geometrical material I will mention only the superb series of books by Martin Gardner, based on his recreational mathematics column in *Scientific American* magazine. Many of these are available cheaply in Penguin editions and will be found in the Penguin catalogue.

    * Starred entries are rather more technical.

BOLD B. *Famous Problems of Geometry and How to Solve Them*, Dover, New York, 1964.

COXETER H. S. M. *Introduction to Geometry*, Wiley, New York, 1961.

* COXETER H. S. M. *Twelve Geometric Essays*, Southern Illinois University Press, 1968.

CUNDY H. M. and ROLLETT A. P. *Mathematical Models*, Tarquin Publications, 1987.

* DAVIS C., GRUNBAUM B. and SCHERK F. A., *The Geometric Vein*, Springer, New York, 1981.

EVES H. *A Survey of Geometry*, vols 1 & 2, Allyn and Bacon, Boston, 1963–5.

* FRANCIS G. K. *A Topological Picture Book*, Springer, New York, 1987.

GRUNBAUM B. and SHEPHERD G. C., *Tilings and Patterns*, W.H.Freeman, San Francisco, 1987.

* HEATH T. L. *The Works of Archimedes*, Dover, New York, n.d.

* HILBERT D. and COHN-VOSSEN S., *Geometry and the Imagination*, Chelsea Publishing Company, New York, 1952.

HILDEBRANDT S. and TROMBA A., *Mathematics and Optimal Form*, Scientific American Library, W. H. Freeman, New York, 1985.

HUNTLEY H. E. *The Divine Proportion*, Dover, New York, 1970.

IVINS W. M. *Art and Geometry*, Dover, New York, 1964.

LINDGREN H. *Recreational Problems in Geometric Dissections and How to Solve Them*, Dover, New York, 1972.

LOCKWOOD E. H. *A Book of Curves*, Cambridge University Press, 1967.

* MANDELBROT B. *The Fractal Geometry of Nature*, W.H.Freeman, San Francisco, 1982.

OGILVY C. S. *Excursions in Geometry*, Oxford University Press, New York, 1969.

PEDOE D. *Geometry and the Liberal Arts*, Penguin, 1976.

STEINHAUS H. *Mathematical Snapshots*, Oxford University Press, New York, 1969.

WENNINGER M. *Polyhedron Models*, Cambridge University Press, 1971.

WEYL H. *Symmetry*, Princeton University Press, 1962.

A

acute-angled triangle dissections What is the smallest number of acute-angled triangles into which an obtuse-angled triangle can be dissected? Mark the incentre of the triangle, D, draw a circle centred on D through the vertex B. Complete the triangles as in the figure, and the dissection is complete in seven pieces.

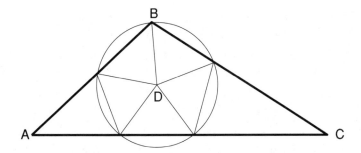

This process works only if $B > 90°$, and $B-A < 90°$ and $B-C < 90°$. If these conditions are not satisfied, then a line can be drawn from B to AC which cuts off one acute-angled triangle and leaves an obtuse-angled triangle which does satisfy the condition, making a total of eight pieces.

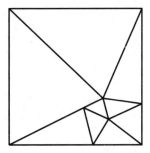

A square can be dissected into nine acute-angled triangles, as the figure above shows, in which several of the angles are close to 90°.

REFERENCES: V. E. HOGGATT, 'Acute isosceles dissection of an obtuse triangle', *American Mathematical Monthly*, November 1961; MARTIN GARDNER, 'Mathematical Games', *Scientific American*, June 1981.

angle in the same segment Mark two fixed points, A and B, on a circle. T is a variable point. The angle ATB is independent of the position of T along the major arc AB. If the variable point is placed at a point on the minor arc AB, call it S, then the angle ASB will be 180° − ATB.

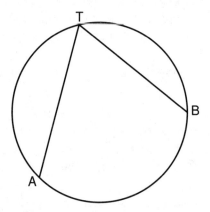

If AB is a diameter, then both angles will be right angles: 'The angle in a semicircle is a right angle', as Thales discovered in about 600 BC and the Babylonians had recognized as early as 2000 BC.

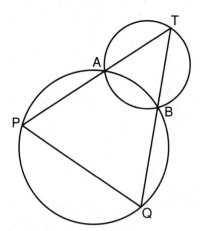

If two circles intersect at A and B, and T moves as before, then the length of the chord PQ is constant.

Regiomontanus posed the question: From what position will a statue such as this appear of maximum size? If the spectator is too close, it will appear heavily foreshortened, but if the spectator is too far away, it will simply be small. The statue subtends the maximum angle at the spectator's eye, and so appears to be of maximum size, when the circle also passes horizontally through the spectator's eye.

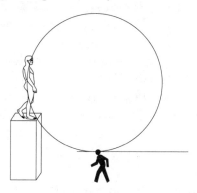

This problem has been rediscovered several times since, most recently in this form: From where should a rugby player take a conversion, which, according to the rules, must be taken from a point in line with the point of touchdown, along a line perpendicular to the goal-line if the try has *not* been scored between the posts?

Apollonian gasket *or* **packing**   When three circles touch each other, they form a curvilinear triangle. Within this triangle another circle touching all three sides can be drawn, forming in turn three curvilinear triangles. This can be repeated over and over again. The figure shows the first few stages of the formation of the Apollonian gasket within this triangle.

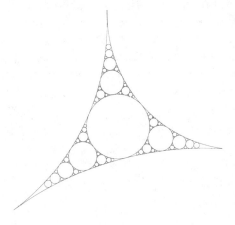

The points that are never inside any of the circles form a set of zero area which is, as it were, more than a line, but less than a surface. Its fractal dimension therefore lies between 1 and 2, though its exact value is not known. It is approximately 1·3.

Apollonius' problem The problem of constructing a circle which will touch each of three given circles was first proposed and solved by Apollonius of Perga. In the most general case, there are 8 solutions: one circle which touches all three without surrounding any of them, one circle which touches and surrounds all three, three circles which surround one of the circles and three which surround two of them. (The analogous three-dimensional problem of finding a sphere to touch each of four given spheres has, in the most general case, $2 \times 2 \times 2 \times 2 = 16$ solutions.)

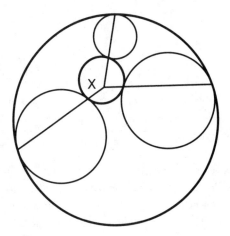

In this figure the inner and outer circles each touch the other three circles, and when the points of contact are joined the three lines are concurrent at X. It follows that any circle which touches the inner and outer circles in the same manner will also have points of contact in line with X.

For every set of four touching circles, there is another set which touch at exactly the same set of six points.

Given the sizes of three circles, each touching both the other two, what is the formula connecting the sizes of the various circles which touch each of them?

The simplest formula uses not the radius of each circle, but its 'bend', which is the reciprocal of the radius.

The French mathematician and philosopher Descartes gave a formula, equivalent to the following, for the bends of four circles touching each other: $2(a^2 + b^2 + c^2 + d^2) = (a + b + c + d)^2$

There is only one formula for eight possible circles, because the bend of a circle can be counted as negative if another circle touches it internally.

This formula was rediscovered in 1842 and again in 1936 by Sir Frederick Soddy, the discoverer of *Soddy's hexlet*. This so pleased him that he celebrated by writing a poem to the journal *Nature*. The middle verse runs:

> Four circles to the kissing come,
> The smaller are the benter.
> The bend is just the inverse of
> The distance from the centre.
> Though their intrigue left Euclid dumb
> There's now no need for rule of thumb.
> Since zero bend's a dead straight line,
> And concave bends have minus sign,
> *The sum of the squares of all four bends*
> *Is half the square of their sum.*

arbelos   This figure, bounded by three semicircles on the same line, was called an *arbelos* (the Greek word for a shoemaker's knife) by Archimedes, who found the radius of a single circle touching all three semicircles.

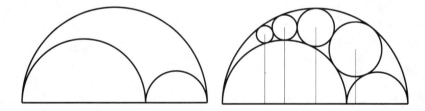

Five hundred years later, Pappus described as an ancient result the fact that if succession of tangent circles are drawn within the arbelos, then the height of the centre of the nth circle above the base-line is n times its diameter.

The centres of the circles lie on an ellipse whose major axis is the base-line, and their mutual points of contact lie on a circle.

Archimedes proved that the area of the arbelos is equal to the area of the circle on the line AC as diameter; adding the other tangent to the two smaller semicircles, BD, gives a rectangle, ABCD.

Archimedes also proved that if two circles are inscribed on either side of AC, touching it, they are equal.

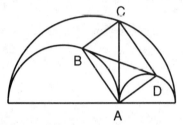 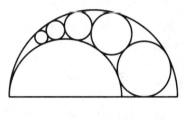

In the figure on the right one semicircle is omitted. Now the distance of the centre of the nth circle from the base is $2n - 1$ times the corresponding radius. Most arbelos figures are special cases of Steiner chains of circles.

Archimedean polyhedra   Archimedes, according to Pappus, investigated the 13 semi-regular polyhedra. Their faces are all regular polygons, but of two or more kinds, and their vertices are identical.

*truncated
tetrahedron*

cuboctahedron

*truncated
cube*

*truncated
octahedron*

*small
rhombicuboctahedron*

*great
rhombicuboctahedron*

*snub
cube*

icosidodecahedron

*truncated
dodecahedron*

*truncated
icosahedron*

*small
rhombicosidodecahedron*

*great
rhombicosidodecahedron*

*snub
dodecahedron*

Eleven of these figures can be obtained by truncation. Nine of these come from truncating the vertices, or the vertices and the edges, of the regular polyhedra. For example, the cuboctahedron is a truncated cube which has been truncated further, until the triangles at the vertices meet at the mid-points of the sides. The others come from truncating two of the first nine.

The snub cube and snub dodecahedron can be constructed by moving the faces of a cube or dodecahedron outwards, giving each face a twist, and filling the resulting space with ribbons of equilateral triangles. Because the twist can be to the left for every face, or to the right, they each exist in two forms which are mirror images of each other.

Archimedean spiral   This curve, which was studied by Archimedes in his book On Spirals, is the locus of a point which moves away from a fixed point with constant speed along a line which rotates with constant velocity.

Its polar equation if the fixed point is at the origin is, therefore, $r = a\theta$. If a > 0, then as the point moves away from the origin it rotates about the origin anticlockwise. If a < 0, the rotation is clockwise.

The Archimedean spiral can be used to trisect any angle, or indeed to divide an arbitrary angle into any given number of equal parts. The angle

XOA is to be trisected. XEFA is a portion of an Archimedean spiral. Draw OB equal to OA, and trisect BX at C and D. Draw arcs from C and D, centre O, to cut the spiral at E and F. Then OE and OF trisect the angle XOA.

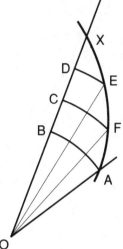

art gallery theorem   In August 1973, at a mathematical conference, Vasek Chvatal asked Victor Klee for an interesting geometrical problem. Klee's response was to ask the novel question: how many guards are necessary to keep all the walls of an art gallery in continuous view?

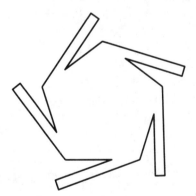

If the art gallery is in the shape of a polygon, with N reflex vertices, then N guards are always sufficient, and sometimes necessary, as the figure shows. A guard is necessary for each of the arms of the gallery. Any one (or more) of these guards can also overlook the central area.

astroid *or* **hypocycloid of four cusps**   The astroid is the locus of a point on a circle rolling inside another circle four times its diameter, and also (as Daniel Bernoulli realized) the locus of a point on a circle, also rolling inside, which has three-quarters the diameter of the fixed circle.

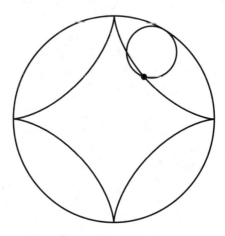

Curiously, in addition to the four visible cusps, it has two imaginary cusps.

If a circle rolls inside another circle of twice its diameter, then the envelope of a diameter of the rolling circle is the astroid. The ends of a diameter of the rolling circle always lie on a pair of perpendicular diameters of the fixed circle, so the astroid is also the envelope of a line of fixed length which slides between two such perpendicular lines.

If the radius of the fixed circle is *a*, then the equation is

$$x^{2/3} + y^{2/3} = a^{2/3}$$

which appears in Leibniz's correspondence in 1715.

The area of the astroid is three-eighths that of its circumscribed circle, or one-and-a-half times that of its inscribed circle.

As the figure below shows, the astroid is the envelope of a family of ellipses whose axes lie on the same pair of perpendicular lines, and for which the sum of the major and minor axes is constant.

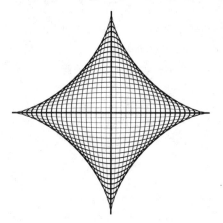

Aubel's theorem Draw any quadrilateral. It need not be convex, and it doesn't even matter if one of the sides is of zero length. Construct squares on all the sides, facing outwards. The line segments joining the centres of opposite squares are equal in length, and perpendicular.

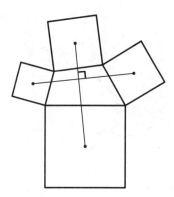

If the squares are constructed in the opposite direction, inwards, then the centres of opposite squares can still be joined by two perpendicular segments of equal length. Moreover, these two shorter segments plus the two in the figure have only two mid-points between them, and the mid-point of the line joining these mid-points is the centroid of the four vertices of the original quadrilateral.

average of two polygons   Draw two similar triangles, in any position but the same orientation. (One of them must not be turned over.)

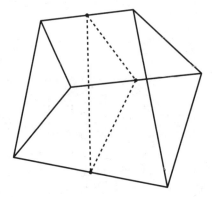

Then the average triangle, formed by joining corresponding vertices and taking the mid-points, is similar to the two original triangles. The same is true of polygons in general. It is also true if, instead of taking the mid-points of the lines, they are just divided in the same ratio.

It is also a special case of this theorem illustrated in this figure:

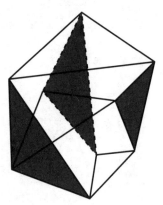

Take two similar triangles in the same orientation. Construct three other triangles, also similar to each other, on the lines joining corresponding vertices. Then the free vertices of these new triangles form a triangle similar to the original pair. The generalized Napoleon figure is also a special case of this theorem.

Another special case, which has been discovered many times, is that if two squares ABCD and XYZD have a common vertex D, then the two mid-points of the lines joining AX and CZ, and the centres of the squares, form another square.

B

Bang's theorem   The faces of a tetrahedron all have the same perimeter only if they are congruent triangles. It is also true that if they all have the same area, then they are congruent triangles.

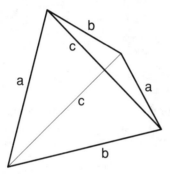

billiard ball path in a cube and in a regular tetrahedron   Can a billiard ball bounce continuously around the inside of a cube always returning after one circuit to its starting point?

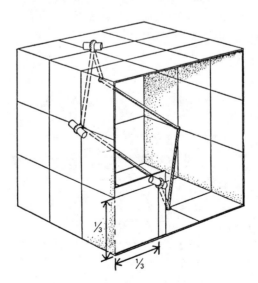

It can. This path was discovered by Hugo Steinhaus. Each bouncing point is the corner of a three-by-three grid on that side, and all the segments of the path are of equal length. The path is known to chemists as a 'chair-shaped hexagon'. Its projection perpendicular to any face of the cube is a rectangle; the projection along one of the diagonals of the cube is a regular hexagon.

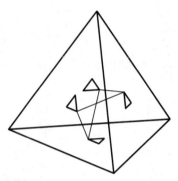

John Conway discovered a similar path inside a regular tetrahedron. The sides of the small triangles marked out on the faces of the tetrahedron are one-tenth the side of the original figure. There are three such paths, one for each corner of a small triangle.

REFERENCE: MARTIN GARDNER, *Sixth Book of Mathematical Games from Scientific American*, W. H. Freeman, San Francisco, 1971.

billiard ball paths in polygons   Can a billiard ball bouncing around inside an acute-angled triangle move in a continuous path? The only closed path of one circuit is the *pedal triangle*, which joins the feet of the altitudes and which is the shortest circuit of any kind which joins the three sides continuously.

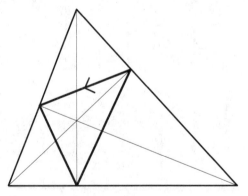

If the ball is allowed to make more than one circuit before returning to its original point and repeating, then an infinite number of circuits are possible, but their segments are all parallel to the sides of the pedal triangle:

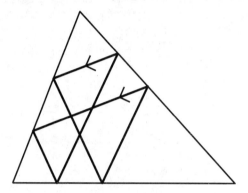

A continuous path is possible inside a quadrilateral if it is cyclic, and if the centre of the circle lies inside the quadrilateral.

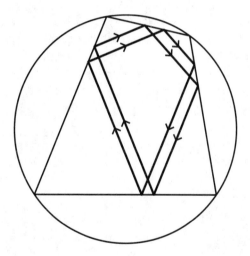

Blanche's dissection   It is a well known and rather difficult problem to dissect a rectangle into squares of different sizes. Turning the problem round, it is easy to dissect a square into rectangles of different sizes, but can the rectangles be of the same area but different shapes?

This is the simplest solution, requiring seven pieces, showing one possible set of dimensions.

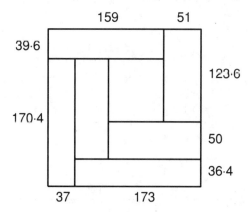

REFERENCE: BLANCHE DESCARTES, 'Division of a square into rectangles', *Eureka*, No.34, 1971.

blancmange curve   Take a series of zigzag curves, each half the height of the previous one and with twice as many zigzags. Continue the series to infinity and then add them all up. The result is the blancmange curve, which is continuous but does not have a tangent anywhere. The first four stages in its construction are shown below. In each figure but the first, the bold line is the sum of the previous stage and the new zigzag.

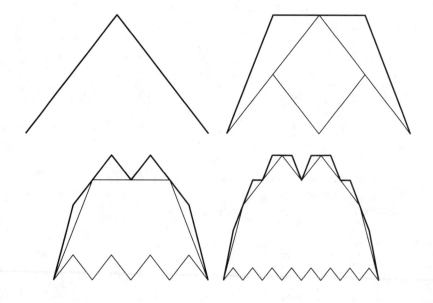

The fifth step shows the blancmange shape more prominently; by the eighteenth step it is difficult to distinguish the curve from its appearance after an infinite number of steps:

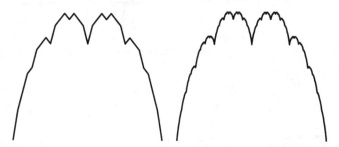

The figure below shows another property of the blancmange curve. Construct one 45° zigzag over two blancmanges, and add them together: the result is a single, larger blancmange.

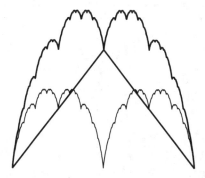

Blashke's theorem The width of a closed convex curve in a given direction is the distance between the two closest parallel lines, perpendicular to that direction, which enclose the curve. The figure shows the widths of three closed convex curves in the given directions.

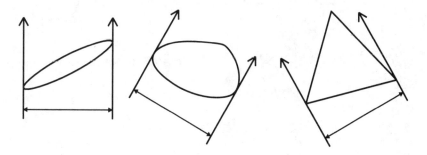

Blashke proved that any closed convex curve whose minimum width is 1 unit or more can contain a circle of diameter 2/3 unit. An equilateral triangle of height 1 unit contains just such a circle, so the limit of 2/3 is the best possible.

Borromean rings   The arms of the Italian Borromean family were three rings, joined together so that all three cannot be separated, although no single pair of rings is linked. The same pattern has been used by the Ballantine Beer Company in the United States and by Krupp, the German armaments manufacturer.

There are no distinct right-handed and left-handed forms – either can be manipulated into the other. This has suggested the following three-dimensional version, which has three planes of symmetry.

It is simple to link any number of rings in the same manner.

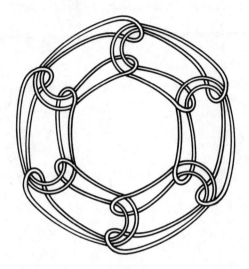

braced square Given a square made of four equal rods, hinged at the corners, how many more rods, of the same length and also hinged at their ends, must be added in the same plane to make the original square rigid in that plane?

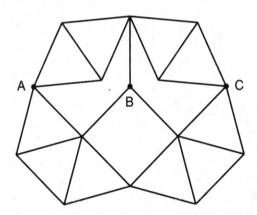

This is the minimum solution, found by readers of Martin Gardner's column in *Scientific American*. The points A, B and C are collinear.
REFERENCE: MARTIN GARDNER, *Sixth Book of Mathematical Games from Scientific American*, W. H. Freeman, San Francisco, 1971.

braids   The best-known braid is the repeating braid used to plait long hair. It comes in two forms, right-handed and left-handed.

If it is stopped at some point and the corresponding ends joined, the result is either three linked rings or a single knot.

Brianchon's theorem   If a hexagon is circumscribed about a conic, that is, if each of its sides touches the conic, then the major diagonals of the hexagon are concurrent.

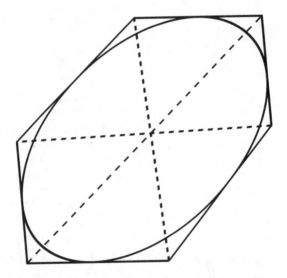

As Brianchon also showed, the sides of the circumscribing hexagon can be taken in any order.

The major diagonals of the hexagon formed by the points of contact meet in pairs on the diagonals of the hexagon.

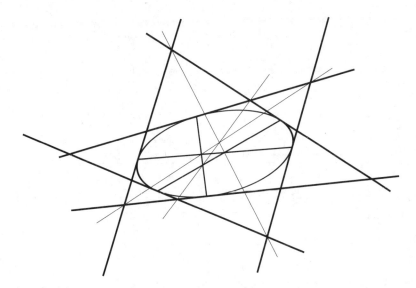

Brianchon published his theorem in 1810. It is the dual of the much earlier Pascal's theorem and can therefore be obtained from Pascal's theorem by switching lines and points, thus:

BRIANCHON'S THEOREM:
If a hexagon is *circumscribed* about a conic – that is, if each of its *sides* touches the conic, then the *lines* joining pairs of opposite *vertices* pass through one *point*.

PASCAL'S THEOREM:
If a hexagon is *inscribed* in a conic – that is, if each of its *vertices* lies on the conic, then the *points* in which pairs of opposite *sides* meet lie on one *line*.

Brocard points of a triangle   Named after Henri Brocard, a French army officer, who described them in 1875. However, they had been studied earlier by Jacobi, and also by Crelle, in 1816, who was led to exclaim, 'It is indeed wonderful that so simple a figure as the triangle is so inexhaustible in properties. How many as yet unknown properties of other figures may there not be?' How prophetic! Entire books have been written on this figure.

For any triangle there is a unique angle ω, the Brocard angle, such that the lines in the figure concur, at the Brocard points Ω and Ω'.

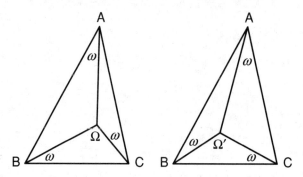

The Brocard angle is given by this formula whose simplicity suggests that it must be significant:

$$\cot \omega = \cot A + \cot B + \cot C$$

The Brocard points can be constructed geometrically by drawing the circles that pass through two vertices, touching one side, as in this next figure. The circles touching AB at A, and so on, define one Brocard point, and the circles touching AB at B, and so on, would define the other.

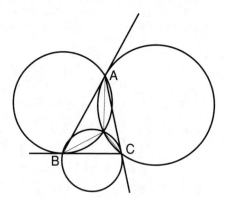

Here are two more 'wonderful' properties: If $C\Omega'$ and $B\Omega$ meet at X and X', and so on, then Ω, Ω', X, Y, and Z all lie on a circle. If three dogs start at the vertices of a triangle and chase each other's tails, each moving at the same speed, then the final dogfight will take place at one or other of the Brocard points, according to the direction of the chase. Compare the fate of four dogs chasing each other, under *pursuit curves*.

C

Cairo tessellation So called because it often appears in the streets of Cairo, and in Islamic decoration.

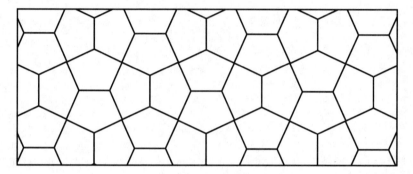

It can be seen in many ways, for example as cross-pieces rotated about the vertices of a square grid, their free ends joined by short segments, or as two identical tessellations of elongated hexagons, overlapping at right angles. The latter suggests that the Cairo tile has many different forms, depending on the shape of the overlapping hexagons.

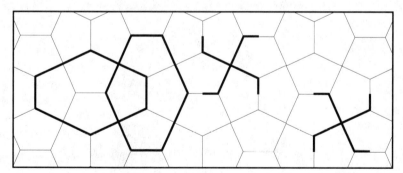

Its dual tessellation, formed by joining the centre of each tile to the centre of every adjacent tile, is a semiregular tessellation of squares and equilateral triangles.

cardioid *or* **epicycloid of one cusp**   The cardioid (meaning 'heart-shaped'), together with related curves such as the astroid, was first studied in 1674 by the astronomer Ole Rømer, who was seeking the best shape for gear teeth. Earlier, the Greeks had considered describing the motion of the planets as 'circles-moving-upon-circles'.

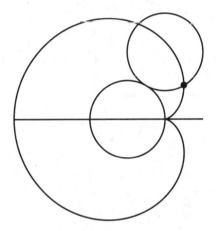

When a circle rolls round another circle of the same size, any point on the first circle traces out a cardioid. Alternatively, it is the path of a point on a moving circle twice the diameter of the fixed circle, which rolls round while enclosing the fixed circle.

The polar equation is $r = 2a(1 \pm \cos\ \theta)$. The length of the cardioid is $16a$, and its area $6\pi a^2$.

The cardioid is also the envelope of all the circles with centres on a fixed circle, passing through one point on the fixed circle.

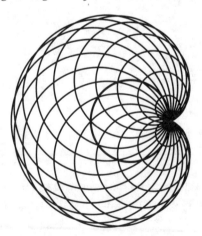

Draw any three parallel tangents, and join the points of contact to the cusp. These three radii are at angles of 120°, and the cusp is a *Fermat point* of the points of contact. The centroid of three points at which parallel tangents touch the cardioid is always the centre of the fixed circle.

An arbitrary line will cut the cardioid in four points, two of which may be imaginary. The sum of the distances from the cusp to these intersections is constant. In particular, since a line through the cusp cuts the curve in two points, the length of any chord through the cusp is constant, and equal to 4*a*. The midpoints of these chords lie on a circle. The tangents at the ends of a chord through the cusp are perpendicular.

carpenter's-square trisection   One of the three classical Greek problems, which cannot be solved by using ruler and compasses alone, is to trisect a general angle. It can be achieved with a carpenter's square.

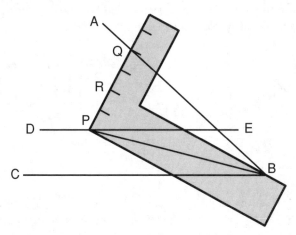

∠ABC is the angle to be trisected. First use the wide arm of the carpenter's square to draw DE parallel to BC. Then lay the square so that one edge goes through B and the outer corner lies on DE, and so that the length PQ is double the width of the wide arm. Mark the mid-point, R, of PQ. Then BP and BR are the trisectors of ∠ABC.

REFERENCE: H. T. SCUDDER, 'How to trisect an angle with a carpenter's square', *American Mathematical Monthly*, May 1928.

Cassinian oval *or* **ellipse**   If a point moves so that the product of its distances from two fixed points, F_1 and F_2, is constant, its path is a Cassinian oval – named after Giovanni Domenico Cassini, who studied them in 1680 in connection with the relative motions of the Earth and the Sun.

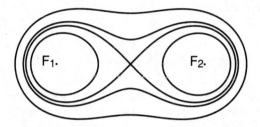

Bernoulli's lemniscate is a special case in which the constant product is equal to the square of the distance between the fixed points.

Cassini's ovals are the cross-sections of a circular torus cut by a plane parallel to its axis. The Greek mathematician Perseus first considered the sections of a torus, so they have been called the spiric sections of Perseus (the Greeks, curiously, having called the torus the *spira*).

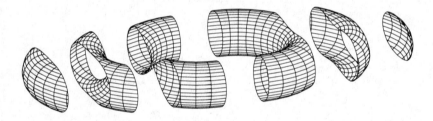

catenary A uniform hanging chain forms a catenary, so named by Huygens in 1691. Galileo thought that a rope might hang in the shape of a parabola, an understandable mistake since the parabola and catenary are very close to each other near the vertex.

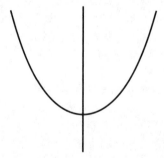

The equation of the catenary is $y = a \cosh(x/a)$.

The catenary is also the locus of the focus of a parabola which rolls on a straight line.

The involute of the vertex of the catenary is the tractrix. The asymptote of this tractrix is called the directrix of the catenary.

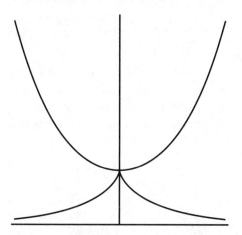

catenoid The surface formed by rotating a catenary about its directrix is a minimal surface. It is the form of a soap film between two empty circular rings on the same axis. It is the only minimal surface of revolution.

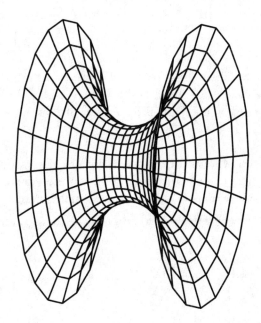

caustic of a circle   Caustics were first studied as a branch of optics by Tschirnhausen in 1682.

Given a fixed curve, and a fixed source of light, the light rays from the source which are reflected from (or are refracted by) the curve, envelope a new curve called a caustic.

The caustic of a circle produced by reflection is seen, rather crudely, when a lamp shines against the inside of a teacup and the light rays are reflected onto the surface of the liquid.

The caustic by reflection is generally a limaçon. There are three exceptional positions for the light source. At infinity, the caustic is a nephroid, if the light source is on the circle it is a cardioid, and the caustic of a light source at the centre of the circle is the centre of the circle itself.

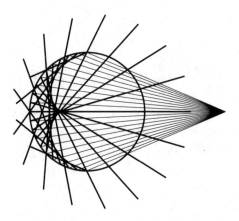

The figure above shows the caustic of a circle, by reflection, for a point source outside the circle. Caustics by reflection can also be thought of as evolutes. The curve in the figure is the evolute of a limaçon whose pole is the light source.

Ceva's theorem   Giovanni Ceva was a geometer and hydraulic engineer, and also the first mathematician to write on economics. In 1678 he published a book containing the theorem named after him, which he proved by considering centres of gravity.

If

$$\frac{BA'}{A'C} \cdot \frac{CB'}{B'A} \cdot \frac{AC'}{C'B} = 1$$

then the lines A A′, BB′ and CC′ concur. The converse is also true. Looked at mechanically, as Ceva viewed it, A′, B′ and C′ are the centres of gravity of suitable pairs of weights at the vertices, and the point of concurrence is the centre of gravity of all three weights together.

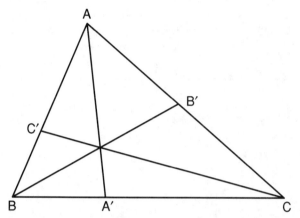

The theorem can be extended to any simple polygon with an odd number of sides. In a pentagon, for example, if lines through the vertices A, B, C, D and E, meet the opposite sides in A′, B′, C′, D′ and E′, then

$$\frac{AC′}{C′E} \cdot \frac{EB′}{B′D} \cdot \frac{DA′}{A′C} \cdot \frac{CE′}{E′B} \cdot \frac{BD′}{D′A} = 1$$

chords at 60° Given any closed convex curve, it is possible to find a point P, and three chords through it inclined at 60°, such that P is the mid-point of all three.

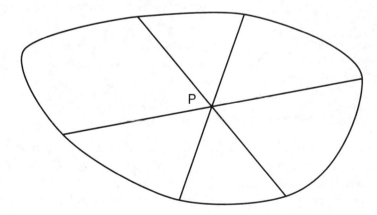

chords halving the perimeter Every diameter of a circle, or every straight line through the centre of a square (or more generally a parallelogram) bisects the perimeter. However, a curve may possess a point with this property without being so symmetrical.

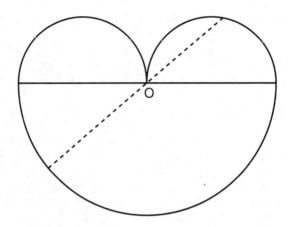

This is two equal half-circles on another half-circle. Every line through O divides the perimeter into two equal parts.

circle tessellations Tessellations are usually defined as filling the plane completely, but the concept is easily extended to tessellations with holes in them, or tessellations of circles, which necessarily leave gaps everywhere.

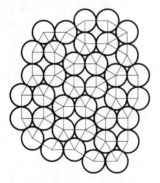 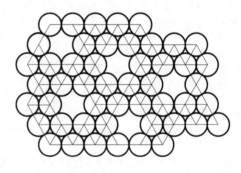

All the semiregular tessellations can be transformed into a network of circles by drawing identical circles centred on every vertex. On the left is the transformed tessellation of squares and equilateral triangles, and on the right the result of transforming one of the two semiregular tessellations of hexagons and equilateral triangles. (That tessellation comes in a right-handed and a left-handed form.)

circles on a sphere   How large can N identical circles be, if they are to be placed on the surface of a sphere? For particular values of N, for example if N is the number of faces of one of the regular polyhedra, the solution is simple, and completely symmetrical. Thus eight identical circles can be drawn, one within each quadrant of the surface of a sphere, each touching three others, corresponding to the faces of an octahedron.

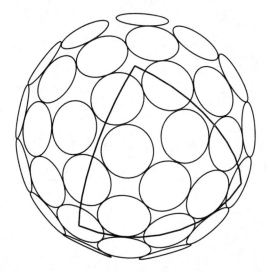

For other values of N, the configuration is less symmetrical and the solution much harder to find. The figure shows a solution for 64 circles. The spherical triangle shows the position of the pole (with four circles surrounding it) and the equator as the side opposite to the pole.

circumcircle of a triangle   The perpendicular bisectors of the sides of a triangle meet in the point which is the centre of the circle through the vertices. If H is the orthocentre of the triangle, then the sum of the vectors of OA, OB and OC is equal to the vector of OH.

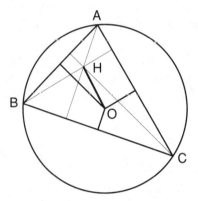

Clifford's theorems   Clifford discovered a sequence of theorems, each building on the last in a natural progression.

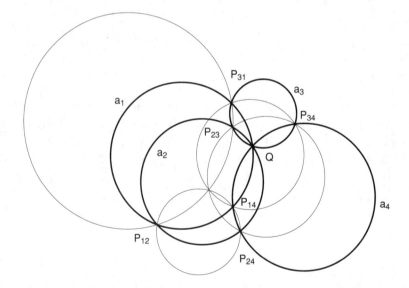

Clifford's first theorem: Let a_1, a_2, a_3 and a_4 be four circles passing through a point Q. Let a_1 and a_2 meet also in P_{12}, and so on. Let a_{123} be the circle through P_{12}, P_{23} and P_{31}, and so on. Then, the four circles a_{123}, a_{124}, a_{134} and a_{234} all pass through one point, P_{1234}.

Clifford's second theorem follows on naturally: Let a_5 be a fifth circle through Q. Then the five points P_{1234}, P_{1235}, P_{1245}, P_{1345} and P_{2345} all lie on a circle a_{12345}.

Clifford's third theorem is: The six circles a_{12345}, a_{12356}, ..., a_{23456} all pass through the point p_{123456}.

This sequence of theorems continues for ever.

coaxial circles   This figure shows two sets of coaxial circles. One set consists of all the circles through two fixed points. Each circle of the second set is orthogonal to every circle of the first set; that is, they cross at right angles.

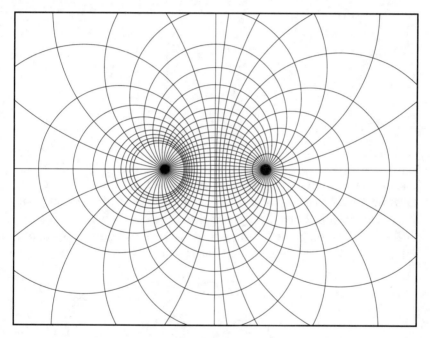

The circles in one set do not meet each other, and they include as limiting cases the two points inside the smallest circle and the vertical line of symmetry, which can be thought of as a circle of infinite diameter.

Every circle in the other set of circles passes through the two limiting points of the first set, and includes the horizontal axis of symmetry as a special case. It has two imaginary limiting points.

The figure was produced by inversion of a set of concentric circles (with radii increasing regularly) together with a set of lines (spaced at equal angles) through their centre. The inverting circle is centred on the left limiting point and has a radius equal to the distance between the two centres. Each of the circles that do not meet is the inverse of one of the concentric circles. Those with centre falling outside the inverting circle give rise to the circles on the left, and so have a different spacing from those on the right. Each intersecting circle in the other set is the inverse of one of the lines.

collapsoids   Jean Pedersen, while typically experimenting with something else, discovered a class of non-convex collapsible polyhedra.

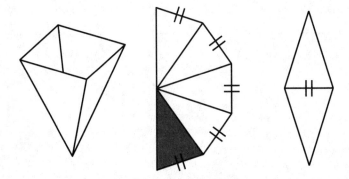

Imagine that each edge of an icosahedron (or dodecahedron – the result is the same) is replaced by a baseless pyramid, the vanished edge being one of its diagonals. Each baseless pyramid has the net shown in the centre above and 30 of them are fitted together using the tabs as shown above right. This gives the following polar collapsoid.

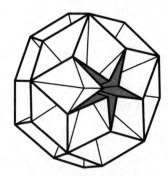

REFERENCE: JEAN PEDERSEN, 'Collapsoids', *Mathematical Gazette*, No. 408, 1975.

common chords Given three intersecting circles, their common chords pass through a common point.

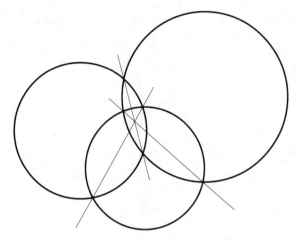

If the circles do not intersect in real points, their common chords are not real, but they still meet in a common real point which is the meet of the three radical axes of the circles. This point is the centre of the unique circle which cuts all three circles orthogonally.

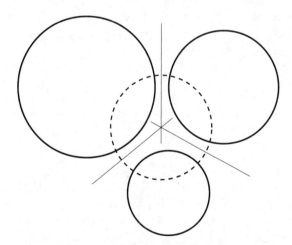

complete quadrilateral Any four general lines meet in six points, forming a complete quadrilateral. A complete quadrilateral has three diagonals, in contrast to an 'ordinary' quadrilateral. The mid-points of these diagonals lie on a straight line.

Newton proved that if a conic is inscribed in a quadrilateral, then its centre lies on the line joining the mid-points of its diagonals.

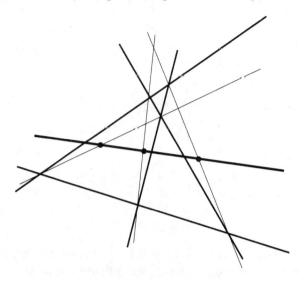

The four lines form four triangles, whose orthocentres lie on a line which is perpendicular to the line formed by the mid-points of the diagonals, and whose circumcircles have a common point.

Plücker proved that the circles on the three diagonals as diameters have two common points. The common points lie on the straight line joining the orthocentres of the four triangles.

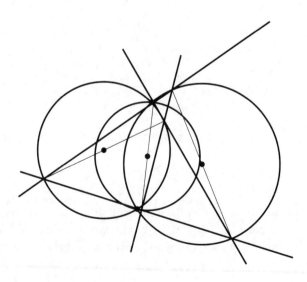

compound polyhedra   Eight vertices of a regular dodecahedron can be chosen to be the vertices of a cube in five different ways. The figure shows two of these cubes placed at the vertices of the dodecahedron. Constructing all these cubes at once produces the compound polyhedron of five cubes in a dodecahedron.

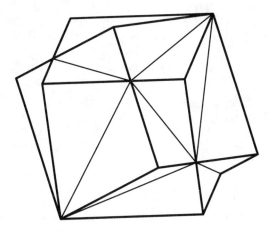

Similarly, five regular tetrahedra can be found in the dodecahedron, to produce a symmetrical compound polyhedron, in two different ways: one left-handed and one right-handed. The twenty faces of the five tetrahedra form, invisibly inside the compound, the faces of an icosahedron whose vertices are the dimples where five edges of the tetrahedra meet.

It is also possible for a pair of dual polyhedra to form a symmetrical compound, because they have the same numbers of edges and the same symmetries. These are the compounds of cube and octahedron (left), and dodecahedron and icosahedron (right):

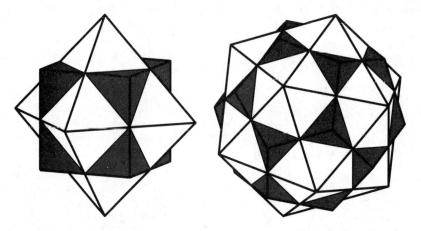

The polyhedron common to the dodecahedron and icosahedron is the icosidodecahedron, obtained by removing the protruding pyramids. The polyhedron which contains them both is the rhombic triacontahedron.

conchoid of Nicomedes   Take any curve and a fixed point, A, not on the curve, and a constant distance k. Draw a straight line through A to meet the curve at Q. If P and P′ are points on the straight line such that P′Q = QP = k, then P and P′ trace out the conchoid of the curve with respect to A.

A practical method is to attach two pens to opposite ends of a ruler, insert a pin at the fixed point, and allow the ruler to move against the pin so that the centre of the ruler moves along the fixed line.

This is the conchoid of the straight line with respect to the point A.

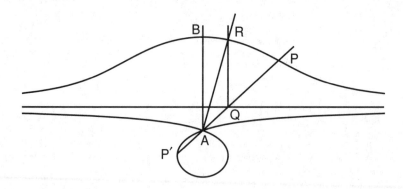

The conchoid of a curve will vary according to the fixed point chosen. Special choices of the fixed point will produce especially simple results. For example, the conchoid of a circle, with respect to a fixed point on the circle, is a limaçon of Pascal.

Nicomedes invented the conchoid ('mussel-shell-shaped'), according to Pappus, in order to solve both the problem of duplicating the cube and the problem of trisecting the angle. This is how it performs the latter feat: in the figure, let $AQ = \frac{1}{2}QP = \frac{1}{2}k$, and let QR be perpendicular to the line. Then $\angle RAB = \frac{1}{3}\angle PAB$.

confocal conics   Given any pair of points, there are an infinite number of ellipses and hyperbolas with these points as foci.

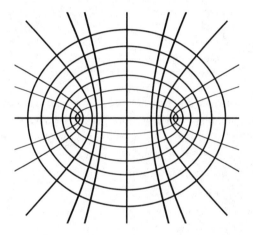

No ellipse meets any other ellipse, nor does any hyperbola meet any other hyperbola, but every ellipse meets every hyperbola and cuts it at right angles.

Given only one point, and a line through it, there are two infinite families of parabolas with the point as focus and the line as axis. Each parabola of one set is orthogonal to every parabola of the other set.

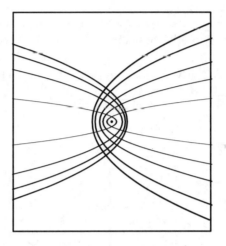

Cremona–Richmond configuration The simplest configurations of points and lines, such as the *Fano plane*, *Desargues's configuration* or the *eleven–three (11₃) configurations*, all contain at least one set of three points and three lines joining them to form a triangle. Indeed, it seems quite natural that any configuration should contain some triangles.

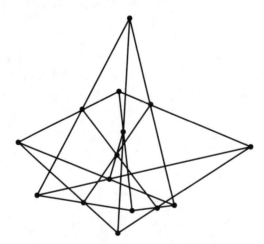

The Cremona–Richmond is a 15_3 configuration, with 15 lines, 15 points, 3 lines through every point, 3 points on every line, and not one triangle.

cross-ratio   Pappus proved in the seventh book of his *Mathematical Collection* that, if four lines through a point are cut by two transversals, then the ratios, called the cross-ratios, of A, B, C and D and A′, B′, C′ and D′ respectively, are equal. The subject of cross-ratios then lay fallow until Desargues developed it in his *Brouillon Project* of 1639.

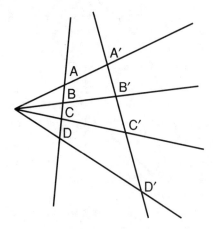

The cross-ratio can be thought of as a ratio of ratios: AB/BC divided by AD/DC. The cross-ratio of four concurrent lines is the cross-ratio created by any line crossing them.

cube   The cube is the best known of the Platonic or regular solids. It has 6 faces, 8 vertices and 12 edges; and 13 axes of symmetry, 3 through the centres of opposite faces, 4 through opposite vertices and 6 through the mid-points of pairs of opposite edges. It is also a zonohedron.

Identical cubes fill space most naturally when each cube meets each of its neighbours across a whole face. However, they can fill space in an infinite number of ways. Not only will layers of cubes slide against each other, but cubes can be arranged in each layer in an infinite number of ways. No other space-filling solid has this flexibility.

Take a cube and delete the edges through a pair of opposite vertices. The mid-points of the remaining edges are the vertices of a plane regular hexagon. If some cubes are stacked to fill space in the natural way, the same plane cut which creates this regular hexagon in one cube will cut the stack in the semiregular tessellation of regular hexagons and equilateral triangles.

There are four ways of bisecting the cube by a cut forming a regular hexagon. The edges of all the hexagons are the twenty-four edges of a cuboctahedron.

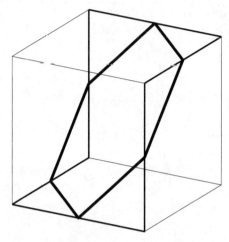

The dual of the cube, formed by joining the centre of each face to the centres of the adjacent faces, is a regular octahedron.

This is a compound of three cubes forming crosses on each other's faces. Each pair of cubes shares one axis of symmetry through a pair of opposite faces.

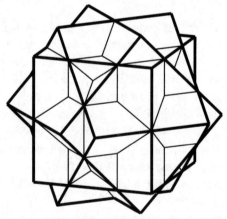

cubic and triangle   In the second half of the nineteenth century and the early part of the twentieth, there was an upsurge of interest on the part of a few mathematicians in what was called the 'Modern Geometry of the Triangle'. Many new features of the triangle were discovered and named, often after the discoverers: the Brocard points, the Gergonne point, Nagel's point, Lemoine points, Tucker's circle, Neuberg's circle, Fuhrmann's circles, Kiepert's hyperbola, and so on.

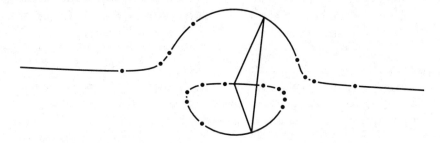

The figure shows one high point of their endeavours – a cubic curve, with one asymptote, which passes through no less than 37 significant points related to a general triangle; 21 are shown in the figure. Among the points lying on the cubic are: the vertices, the reflections of the vertices in the opposite sides, the six vertices of the equilateral triangles constructed outwards and inwards on the sides, the circumcentre and the orthocentre, and the centres of the inscribed and escribed circles.

The tangents to the cubic at these last four points are all parallel to the asymptote. Among other properties, any line through one vertex cuts the cubic in two points which lie on a circle through the other two vertices.

cyclic quadrilateral A quadrilateral inscribed in a circle. If ABCD is a quadrilateral inscribed in a circle, then angles $A + C = B + D = 180°$.

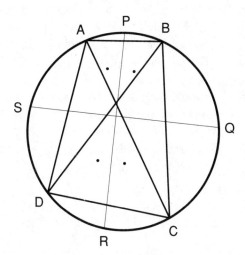

Omit each of the vertices in turn, to obtain the four triangles BCD, ACD, ABD and ABC. The dots in the figure mark the incentres of these four triangles. They form a rectangle.

If P, Q, R and S are the mid-points of the arcs AB, BC, CD and DA, then the sides of the rectangle are parallel to PR and QS, and PR and QS meet at the centre of the rectangle.

If the excentres of the same four triangles are added, then together with the incentres, they form a rectangular 4 × 4 grid of 16 points.

The centroids of the same four triangles form a quadrilateral similar to the original, as do their four nine-point centres. The four orthocentres form a quadrilateral congruent to the original.

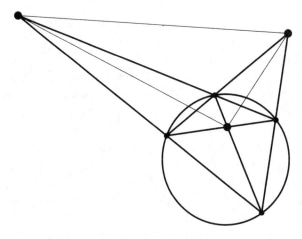

Take four points on a circle and draw all six lines joining them. The three diagonal points form the diagonal triangle (shown as thin lines in the figure). Each vertex is the pole of the opposite side with respect to the circle. If the tangents at all four original points are drawn, they meet in pairs on the sides of the diagonal triangle.

cycloid Marin Mersenne considered problems about the cycloid, but, as was his custom, he passed the problems on to his fellow mathematicians and correspondents. The first treatise on the cycloid was written by Evangelista Torricelli, a student of Galileo, in 1644. Pascal also studied the curve, even using his study to relieve a bad toothache.

When a wheel rolls along a straight surface, a point on the wheel's rim traces a cycloid:

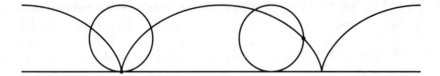

Points within the wheel trace a *curtate cycloid*:

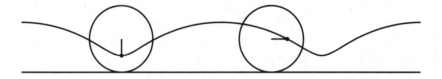

When the wheel of a train rolls along a rail, a point on its circumference traces a *prolate cycloid* which contains loops:

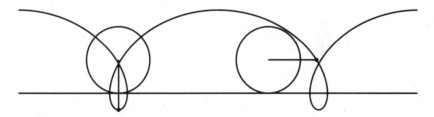

Imagine a circle, twice the diameter of the original circle, rolling with it. Then the diameter of the larger circle which was originally vertical touches the cycloid, which is its envelope.

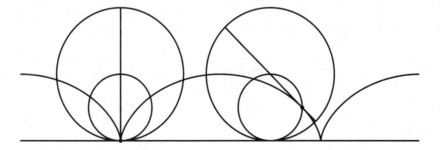

Galileo supposed, correctly, that the cycloid is the strongest shape for the arch of a bridge.

Galileo also attempted to find the area of the cycloid in 1599. Following Archimedes' example, he cut out one complete cycloidal arch, weighed it, and compared the weight with the weight of the generating circle. He concluded that its area is roughly three times the area of the generating circle. Roberval proved in 1634 that it is indeed exactly three times the generating circle in area.

The length of one complete arc equals the perimeter of a square circumscribed about the generating circle, as Sir Christopher Wren, an excellent geometer, proved in 1658.

The evolute of a cycloid is an equal cycloid which is one half revolution out of phase with the original cycloid.

The cycloid is also the solution to the brachistochrone problem: What is the shape of the brachistochrone, the curve down which a particle, falling under gravity, will travel from A to B in the shortest time?

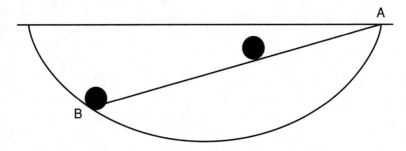

An extraordinary feature of this solution is that if the destination point is only slightly below the height of the starting point, the quickest route takes the particle below the final point, and then up towards it!

It is also true that a particle rolling down a cycloidal groove, provided the axis of the cycloid is vertical, will reach the bottom in the same time whatever point on the cycloid it may have started from. In other words, as well as being brachistochrone, the cycloid is a tautochrone.

Galileo discovered that the period of a pendulum depends only on its length, but this is true only for small oscillations. By making the pendulum wrap round a cycloid, it becomes true for oscillations of any amplitude.

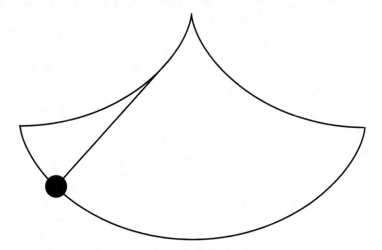

Huygens was the first to use this principle in an attempt to improve the pendulum clock, but the idea created more problems than it solved, and was soon abandoned.

D

Dandelin spheres   An ellipse is a plane section of a cone. It is possible to fit one sphere into the cone to touch the plane, between the plane and the vertex, and another sphere to touch the plane and the cone on the other side.

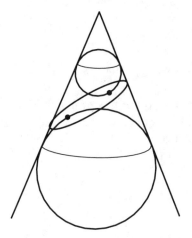

Dandelin, a professor of mechanics at Liège University, proved that the two spheres touch the ellipse at its foci, and that the directrices of the ellipse are the lines in which the cutting plane meets the planes of the circles in which the spheres touch the cone.

degenerate quartics   Any two conics taken together can be treated as a quartic, a curve of degree four. So can a cubic and a straight line.

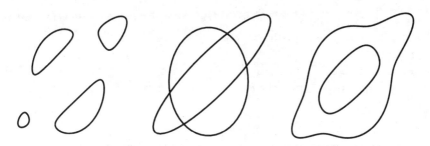

The equation for the quartic is found by taking two ellipses, with equations $E_1 = 0$ and $E_2 = 0$, and forming the equation $E_1E_2 = 0$. If the coefficients of the terms in the equation of the quartic are then varied slightly, the result will be a quartic which is very close to both ellipses. Depending on how the coefficients are varied, there are two possibilities, either four beans, or a curve in only two separate parts.

Every quartic has 28 bitangents, but most of them are usually imaginary. If the coefficients are suitably chosen, then each of the 4 individual beans has 1 bitangent and each of the 6 pairs of beans has 4 bitangents so that all 28 are real.

Delian problem of duplicating the cube   When the Athenians were suffering from a plague in 430 BC, they consulted the oracle of the god Apollo at Delos, and were instructed to double the size of their altar, which was a cube. They at once doubled every edge, and the ravages of the plague increased.

The problem of constructing a length $\sqrt[3]{2}$ times the length they required became known as the Delian problem, although equally ancient, similar problems on the size of altars had been studied in India.

It was soon realized that the problem was equivalent to finding two *mean proportionals* between two lengths. In other words, given a and b, if two mean proportionals x and y can be found such that $x/a = y/x = b/y$, then $(x/a)^3 = b/a$.

Unfortunately, the Greeks were unable to construct solutions by using ruler and compasses only. Their many solutions were obtained by using either operations that can be performed only with an element of human judgement, or curves invented for the purpose (and these curves could not themselves be constructed by ruler and compasses). One such curve is the conchoid of Nicomedes; another is the cissoid of Diocles.

In general, a cissoid can be constructed for any two curves and a given fixed point. The cissoid of Diocles is the cissoid of a circle (centre O) and a line touching it (at B) with respect to the point (A) opposite the point of tangency. Draw a straight line through A to meet the circle at Q and the line through B at R, and mark the point P on this new line such that $AP = QR$. The cissoid is then the path of P. If the radius of the circle is unity, then $OU^3 = OL$.

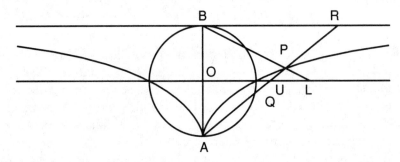

A simple solution to the Delian problem, requiring only a ruler with two points marked on it 1 unit apart, is the following. The unit lengths are as marked. The ruler is adjusted by hand so that it passes through the upper vertex of the equilateral triangle, and the distance between the points where it intercepts the two lines on the right is 1 unit. The distance from the upper vertex to the nearest of the intercepts is then $\sqrt[3]{2}$ units.

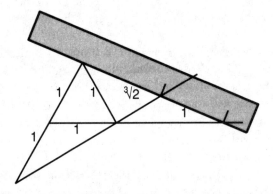

REFERENCES: E. H. LOCKWOOD, *A Book of Curves*, Cambridge University Press, Cambridge, 1961; H. DORRIE, *One Hundred Great Problems of Elementary Mathematics*, Dover, New York, 1963.

deltahedra   Martyn Cundy gave the name 'deltahedron' to any polyhedron whose faces are all equilateral triangles. Three of the Platonic solids are deltahedra: the tetrahedron, octahedron and icosahedron. There are just eight convex deltahedra, the Platonic solids just mentioned, and the five shown below, drawn to show how they can be assembled from smaller parts.

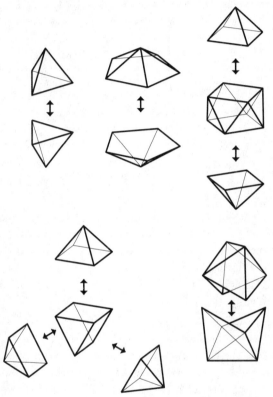

If the solid need not be convex, there are endless possibilities, not least because adding a regular tetrahedron to any face produces a new deltahedron (which, by this definition, is allowed to intersect itself).

An infinite pile of octahedra form an infinite deltahedron. An octahedron can be thought of as a triangular *antiprism*: two equilateral triangles face each other, each vertex of one opposite an edge of the other, and the space between filled in by 2 × 3 = 6 equilateral triangles.

Any two polygons with the same number of edges can be opposite faces of an antiprism, and an infinite pile of them has a cylinder-like surface composed only of triangles.

deltoid *or* **hypocycloid of three cusps**   The deltoid was first studied by Euler in 1745. A circle rolls inside a fixed circle. If the rolling circle is either one-third or two-thirds the diameter of the fixed circle, a point on it traces a deltoid.

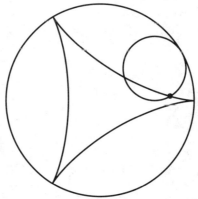

The diameter of a circle of radius two-thirds rolling round a circle of unit radius envelops a deltoid.

Another construction as an envelope is this. Mark a series of numbered points clockwise round a circle, and another set, from the same starting point but double spaced and anticlockwise. Join corresponding points, and the envelope is a deltoid.

A third construction is to take any triangle and draw all its Simson lines. Their envelope is a deltoid.

Let the tangent at T meet the deltoid again at A and B. The length AB is constant and twice the diameter of the inscribed circle, and the mid-point of AB lies on the inscribed circle. The tangents at A and B are perpendicular and meet on the inscribed circle, at the point diametrically opposite to the mid-point of AB, and the normals at T, A and B all meet on the outer circle, at its point of contact with the rolling circle.

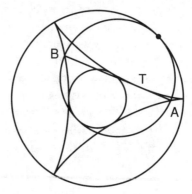

derived polygons   Take any polygon with an even number of sides and join the mid-points of the sides, in sequence. Repeat. The shape tends to a polygon whose opposite sides are parallel and equal in length. The original polygon and all the derived polygons have the same centre of gravity. Alternate polygons are approximately the same shape.

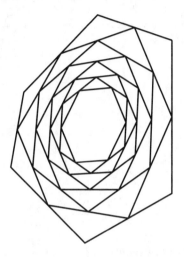

If the sides are divided in a different ratio, not 1 : 1, the same phenomenon occurs, although the derived polygons will not alternate so simply.

If the original polygon is not even plane, but skew, the process nevertheless leads to a plane polygon, with the same property and the same centre of gravity.

Take any hexagon, and find the centres of gravity of each set of three consecutive vertices. These immediately form a hexagon whose opposite sides are equal and parallel in pairs:

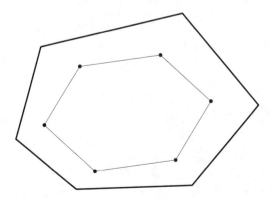

On the other hand, if you take any three consecutive vertices of a hexagon and mark the fourth vertex of the parallelogram of which they are the vertices, the result is the outline of a prism:

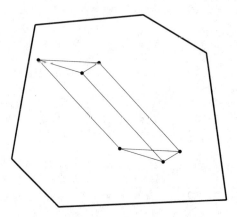

REFERENCE: J. H. CADWELL, *Topics in Recreational Mathematics*, Cambridge University Press, Cambridge, 1966.

Desargues's configuration   Take two triangles which are 'in perspective': that is, the lines joining corresponding vertices pass through a point. Then pairs of corresponding sides meet in three points which are collinear.

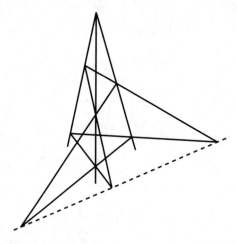

Desargues's theorem, as this is called, can be proved by thinking of it as an essentially three-dimensional figure. The planes ABC and DEF will meet in a line, L. Planes ABC and ABED already meet in the line AB, and planes DEF and ABED already meet in the line DE. Therefore all three of these lines meet at P, the common point of the three planes, which lies on L. Similarly, A C and D F meet at R, on L, and C B and F E meet at Q, on L. When the three-dimensional figure is projected onto the plane, L remains a straight line.

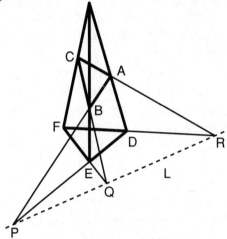

The figure appears to be unsymmetrical, because of the special role of the dashed lines in the explanation. However, this is an illusion. In fact, any point in the figure can be taken to be the special vertex (corresponding to X), and there will then be exactly three labelled intersections in the figure which do not lie on any of the straight lines through it: these three intersections will themselves lie on a line corresponding to PQR in the figure.

The converse of Desargues's theorem is also true: if the meets of pairs of corresponding sides of two triangles lie on a straight line, then the lines joining pairs of corresponding vertices pass through a point.

Moreover, this converse is also the *dual* of the original theorem. In other words, it can be obtained by switching 'point' and 'line', 'line through two points' and 'meet of two lines', in the statement of the original theorem.

dodecagon dissected   Here are two simple and natural ways to dissect a regular dodecagon into rhombuses.

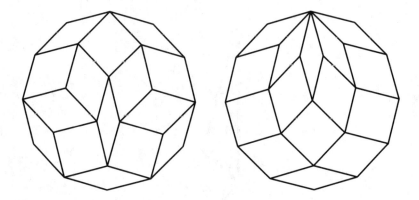

There are three shapes of rhombus in each figure, and although there are several ways of dissecting the polygon into these basic shapes, the proportions of each shape are always the same: 6 narrow and 6 medium rhombuses, and 3 squares.

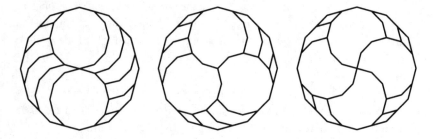

These shapes can be used to construct larger copies of the same shape. In each of the above figures, four dodecagons are dissected into one large copy. Many of the rhombuses remain attached to each other in strips known as *zones*.

The next dissection uses pieces of only one shape, which is an equilateral triangle joined to half a square. The bordered dodecagon has side $\sqrt{2}$ times the original, and twice its area; the larger dodecagon has sides twice that of the original and four times its area. The bordered dodecagon can be extended, using the same piece, to tile the whole plane.

dodecahedron Dodecahedra have 12 faces, and therefore include the regular dodecahedron, with 12 regular pentagonal faces, and the *rhombic dodecahedron*, with 12 rhombic faces.

The regular dodecahedron has 31 axes of symmetry: 10 are threefold, passing through pairs of opposite vertices; 6 are fivefold, passing through the centres of opposite faces; and 15 are twofold, passing through the mid-points of opposite sides.

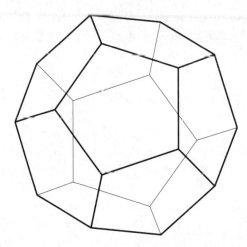

The regular icosahedron has the same number of axes of symmetry, but with 'vertices' and 'faces' reversed in their description.

The relationship between the dodecahedron and the cube can be seen either by joining the mid-points of faces to form the vertices of three rectangles (whose edges are in the golden ratio) which are mutually perpendicular, or by choosing eight vertices of the dodecahedron which are the also the vertices of a cube:

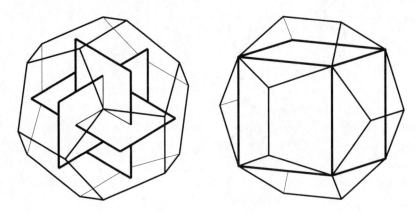

A perhaps surprising fact is that when a regular dodecahedron and a regular icosahedron are inscribed in the same sphere, the dodecahedron occupies a larger proportion of the sphere's volume. The icosahedron has more faces, but the faces of the dodecahedron are more nearly circular.

dragon curve Fold a long strip of paper, right half over left, and open it out to a right angle. Viewed edge-on, this is the dragon curve of the first order. Now, close the strip and fold it in half again, in the same direction as the first fold, and open it out again so that each fold is a right angle. Repeat this process. The results, again viewed edge on, are the dragon curves of the second and third orders. This is the dragon curve of the tenth order:

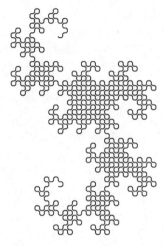

Four dragon curves will fit together around a point, as the next figure of four sixth-order dragons illustrates. In each case, the angles have been slightly adjusted to show that the curve never actually crosses itself and so that you can see the individual curves.

dual polyhedra The dual of any of the Platonic polyhedra is formed by joining the centres of adjacent faces. In the resulting *dual* solid, each vertex corresponds to a face of the original, each face of the new solid to an original vertex, and the edges match, one for one.

As it happens, the dual of each Platonic solid is also Platonic. The regular tetrahedron is its own dual, the cube and the regular octahedron are duals of each other, and so are the regular dodecahedron and icosahedron.

The same simple process will not work for the semi-regular or Archimedean polyhedra, because the centres of the faces round a vertex will not lie in a plane. It is necessary instead to inscribe the semi-regular polyhedron in a sphere and construct the tangent plane at each vertex.

The resulting duals of the semi-regular polyhedra are not themselves semi-regular. However, their faces are all congruent and every vertex is regular, though not all faces are necessarily identical.

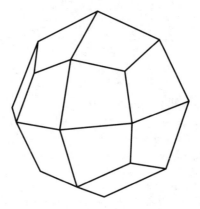

The figure shows the trapezoidal icositetrahedron which is the dual of the small rhombicuboctahedron. The rhombic dodecahedron is the dual of the cuboctahedron. With the rhombic triacontahedron, a *zonohedron*, which is the dual of the icosidodecahedron, it is the only Archimedean dual with rhombic faces.

duals of the semiregular tessellations Every tessellation of regular polygons has a dual, formed by taking the centre of each tile as a vertex of the dual tessellation, and joining the centres of adjacent tiles.

Of the three regular tessellations, that of regular hexagons and that of equilateral triangles are duals of each other, and the tessellation of squares is its own dual.

The semiregular tessellations each have duals which are less regular. Thus the dual of the tessellation of squares and equilateral triangles is the *Cairo tessellation*.

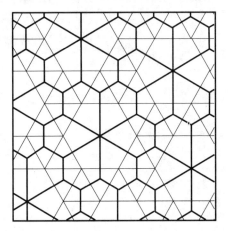

The thick lines show the dual of one of the tessellations of regular hexagons and equilateral triangles.

Dudeney's hinged square-into-equilateral-triangle   Henry Ernest Dudeney exploited dissections in many of his puzzles. This is his masterpiece. Rotate the hinged pieces one way to get the square, and the other way to get the equilateral triangle. Two of the hinges bisect two of the triangle's sides, while the third hinge and the meet of the vertices of two pieces divide the third side in the approximate ratio 0·982 : 2 : 1·018.

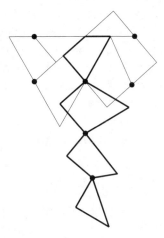

Dudeney made a beautiful wooden model of this dissection, which he was invited to demonstrate before the Royal Society in 1905, an extraordinary but appropriate honour for a master-puzzler.

Dupin cyclide   All the spheres that touch three fixed spheres (each in an assigned manner, either externally or internally) form a continuous chain whose envelope is a Dupin cyclide.

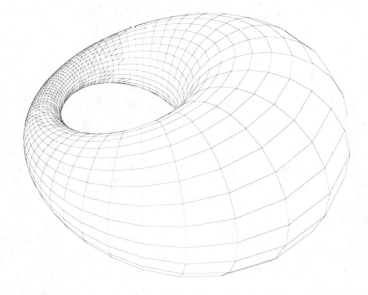

The centres of all the tangent spheres lie on a conic, so an alternative definition of a Dupin cyclide is the envelope of all spheres having their centres on a given conic and touching a given sphere.

A third definition is as the envelope of spheres with their centres on a given sphere and cutting a given sphere orthogonally.

A torus is a special case of a Dupin cyclide, and also, surprisingly, every Dupin cyclide is the inverse of a torus.

E

eleven–three configurations   There are 31 essentially different 11_3 configurations. In each, there are 11 lines and 11 points, with 3 lines through every point and 3 points on every line. These are three of them:

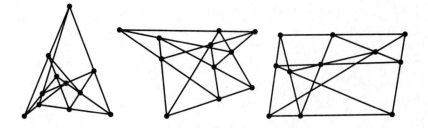

REFERENCE: W. PAGE and H. L. DORWART, 'Numerical patterns and geometrical configurations', *Mathematics Magazine*, March 1984.

ellipse   An ellipse is a plane section of a cone. If the cone is thought of as double, extending on both sides of its vertex, then the plane of the ellipse cuts only one half of the cone. The plane of a cut which produces a parabola is parallel to a line in the surface of the cone, through the vertex, and the plane of a cut producing a hyperbola cuts both halves of the cone.

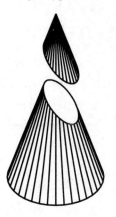

An ellipse can be drawn by fixing a loop of string around two pins, F and G, and a pencil. The path of the pencil as it moves, keeping the string taut, will be an ellipse. F and G are the foci of the ellipse.

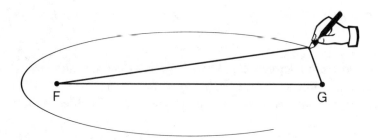

An ellipse also has two *directrices*, one for each focus. An ellipse can be defined as the path of a point which moves so that the ratio of its distance from a fixed point, the focus, to its distance from a fixed straight line, the directrix, is constant and less than one.

If, instead of using two pins, the string is wrapped round another ellipse, the path of the pencil will still trace out an ellipse, with the same foci as the original ellipse.

To draw an ellipse in a rectangle, divide one half of each of the sides and one half of the line joining the mid-point of a pair of opposite sides into an even number of parts, and find the intersections of the lines joining X and Y to the marked points.

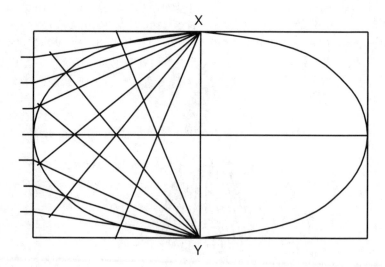

An ellipse can be thought of as a squashed circle. The figure below shows the construction of an ellipse which is the outer circle reduced in height using a factor of 0·6 or alternatively, the inner circle stretched horizontally.

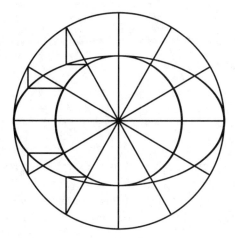

To paper-fold an ellipse, draw a circle and mark a point inside it. Fold the paper so that the circumference falls on the marked point, and crease firmly. Repeat, using different folds. The creases will envelope an ellipse.

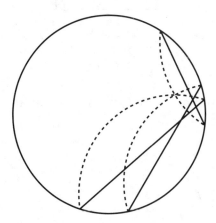

The following method of drawing an ellipse was discovered by Leonardo da Vinci. Cut out a triangle ABC. Draw two axes, which need not be perpendicular, on a piece of paper, and move the triangle so that one vertex moves along one line and another moves along the second line. The path of the third vertex will be an ellipse.

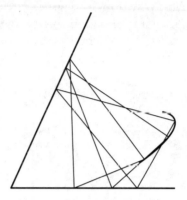

A special case of this construction occurs when a ladder slips against a wall. Any point on the ladder, such as the foot of a person still standing on it, will move in a portion of an ellipse. This is the basis of a commercial instrument for drawing an ellipse using trammels. Two points of a rod slide in two grooves, and the path of a point on the rod is an ellipse.

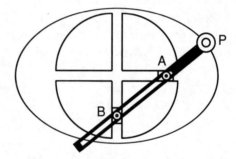

The tangent to an ellipse makes equal angles with the lines joining the point of contact to the foci.

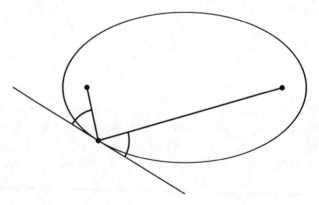

This can be inferred mechanically by considering a small weight sliding on a string attached to two pins. The path of the weight is an ellipse, by definition. At its lowest point the tangent will be horizontal, and provided the weight slides smoothly on the string, the angles of the string to the horizontal will be equal because equal tensions are required if the weight is not moving. So, the tangent makes equal angles with the lines joining the point of contact to the foci.

equal incircles theorem The rays from X are chosen so that the triangles XAB, XBC, XCD, and so on, all have equal incircles. Then the triangles XAC, XBD, and so on also have equal incircles.

Similarly, triangles XAD, XBE, and so on will also have equal incircles, as will triangles XAE and XBF.

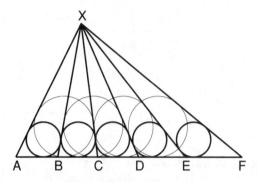

equiangular *or* **logarithmic spiral** Discovered by Descartes in 1638, it cuts any radius through the origin at the same angle. If that angle is called ρ, then the polar equation of the spiral is $r = a \exp(\theta \cot \rho)$.

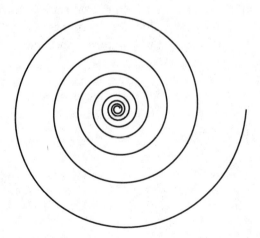

It was studied by Jakob Bernoulli, who was so impressed by its tendency to appear as transformations of itself that he left instructions that the curve be engraved on his tomb, together with the words *Eadem mutata resurgo* ('I shall arise the same though changed').

Its evolute is an equal equiangular spiral, and so is its inverse with respect to the origin. If a light source is placed at the origin, then its caustics by reflection and by refraction are also identical equiangular spirals.

It is similar to itself, in the sense that if any part of the curve is blown up or reduced, it is identical to another portion of the same curve.

If the spiral is rolled along a straight line, then the path of the origin of the spiral, called its pole, is another straight line. The length of the curve from the pole (call it point O) to the point X, is equal to XT, where T is the starting point of the pole and TOX is a right angle.

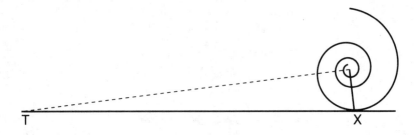

The equiangular spiral occurs again and again in nature. For example, the whorls of the nautilus shell are equiangular spirals. However, patterns such as those in sunflower heads are only approximately equiangular spirals; they are better described by *Fermat spirals*.

equilateral triangle tilings   One of the regular tessellations is composed of identical equilateral triangles. Because the triangles in that tessellation form strips, there are in fact an infinite number of tessellations of the plane composed of the same triangles, but of a less regular nature.

If the triangles can be of several sizes, there are many more possibilities. The following figure shows three different sizes of equilateral triangle tessellating.

Euler line   In any triangle, the circumcentre O, the orthocentre H, and G, the meet of the medians, lie on a straight line. In addition, GH = 2OG. Leonhard Euler published this celebrated theorem in 1765.

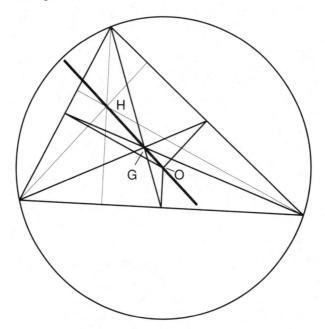

eyeball theorem   The tangents to each of two circles from the centre of the other are drawn. Then the lines AB and XY are equal in length.

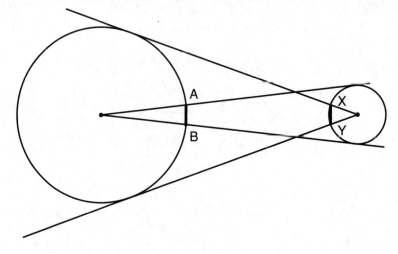

F

face-regular polyhedra   Many polyhedra can be constructed whose faces are regular polygons, but which have little or no other symmetry.

There are five triangles round each vertex of a regular icosahedron, forming a shallow pentagonal pyramid. Slice off three such pyramids and replace them by regular pentagons, and the result is the tridiminished icosahedron.

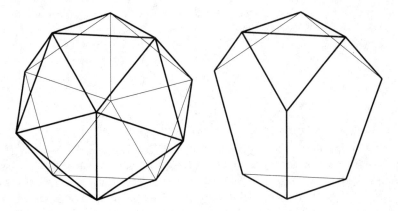

The figure below is known as bilunabirotunda.

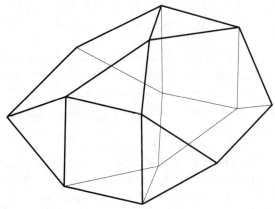

Viktor Zalgaller proved, in 1966, that apart from the regular and semiregular polyhedra and the regular prisms and antiprisms, there are just 92 *convex polyhedra* with regular faces. He named them all, including the gyrofastigium, metabidiminished rhombicosidodecahedron and hebesphenomegacorona. Of the 92, twenty-eight are *simple* in the sense that they cannot be cut into two other face-regular polyhedra.

Fano plane   A finite projective plane consists of points and lines, with the same number of lines through every point and the same number of points on every line.

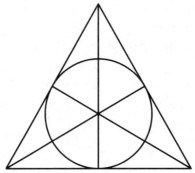

The figure shows the smallest finite projective plane, the Fano plane, which contains 7 points and 7 lines, with 3 points on every line and 3 lines through every point, and is therefore denoted by 7_3. It illustrates the fact that not all finite projective planes can actually be drawn using geometrically straight lines. The Fano plane can at best be drawn geometrically so that all the lines but one are actually straight; the circle is the seventh 'line'.

The total number of points in a finite projective plane is necessarily $1 + p^n + p^{2n}$, where p is a prime number; there will be $1 + p^n$ points on every line, and $1 + p^n$ lines through every point. For the Fano plane, $p = 2$ and $n = 1$.

The Fano plane is the only 7_3 configuration. There is also only one 8_3 which can also be drawn with all but one of the lines geometrically straight. There are three 9_3 configurations, ten different 10_3 ones, thirty-one 11_3 ones and two hundred and twenty-eight 12_3 configurations.

Fatou dust   When the point which generates a Julia set is chosen from outside the *Mandelbrot set* (or the equivalent set for a different transformation), the *Julia set* breaks down into a set of isolated points, called Fatou dust after Pierre Fatou, who worked with Gaston Julia.

If the point is relatively near the boundary of the Mandelbrot set, the Fatou dust is thick, and resembles the Julia sets for nearby points within the Mandelbrot set. As the point moves further and further away from the Mandelbrot set, the dust becomes thinner and thinner.

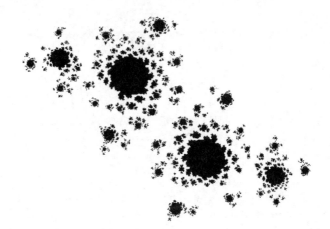

fault-free rectangles   A dissection of a rectangle into several smaller rectangles may include a straight line, called a fault, joining two sides, which divides the original rectangle into two smaller rectangles. Dissections which do not include such lines are called fault-free. A division into 3, 4 or 6 pieces cannot be fault-free. The figure shows a fault-free division into 5 parts and a fault-free division of a 5 by 6 rectangle into fifteen 2 by 1 rectangles.

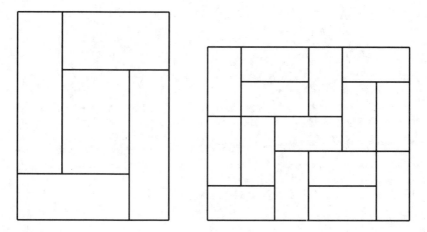

Fermat *or* **parabolic spiral**   Named after Pierre de Fermat, who studied it in 1636; it is alternatively called 'parabolic' because its polar equation is $r^2 = a^2\theta$, which superficially resembles the equation for the parabola: $y^2 = ax$.

Robert Dixon explains how Fermat spirals form more accurate models of the form of plant growth, for example the head of a daisy, than the usual explanation based on the equiangular spiral: the property of the Fermat spiral which is relevant to constructing daisies is that successive whorls enclose equal increments of area.

This is a daisy head constructed on the basis of Fermat spirals:

REFERENCES: R. DIXON, 'The mathematics and computer graphics of spirals in plants', *Leonardo*, Vol. 16, No. 2, 1983; R. DIXON, *Mathographics*, Basil Blackwell, Oxford, 1987.

Fermat point of a triangle   Fermat challenged Torricelli to find the point whose total sum of distances from the vertices of a triangle is a minimum. The problem is quite practical, since if there were three villages at the corners of the triangle, it amounts to asking for the shortest length of road which you would need to build to join all the villages.

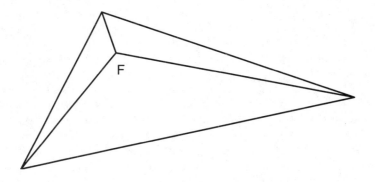

If all the angles of the triangle are less than 120° the desired point, the Fermat point F, is such that the lines joining it to the vertices meet at 120°. If the angle at one vertex is at least 120°, then the Fermat point coincides with that vertex.

The Fermat point can be found by experiment. Let three equal weights hang on strings passing through holes at the vertices of the triangle, the strings being knotted at one point. The knot will move to the Fermat point.

Alternatively, construct an equilateral triangle on each side of the triangle. Then the three lines joining the free vertices of each new triangle to the opposite vertex of the original triangle will all pass through the Fermat point, which is also the common point of the circumcircles of the equilateral triangles (see the figure on the next page). Moreover, these three lines are all of equal length and each equal to the total length of the road network.

If equilateral triangles are drawn on each side facing inwards as in a variant of *Napoleon's theorem* (as shown in the figure on the next page), then the lines joining their free vertices to the opposite vertices of the original triangle (ABC) also meet at a point, P. This point has an extremal property: if the angle at C is less than 60°, and the angles at A and B are both greater than 60°, then PA + PB − PC is a minimum at that point. If the condition is not satisfied, the minimum is attained at either A or B.

If the sides of the triangle are of equal length to a, b, and c, and the distances of the Fermat point from the vertices are x, y and z, then there is a point inside an equilateral triangle of side $x + y + z$ whose distances from the vertices are a, b and c.

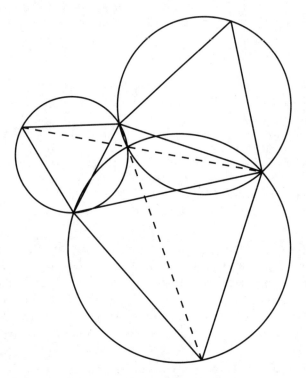

REFERENCE: DAVID NELSON, 'Napoleon revisited', *Mathematical Gazette*, No. 404, 1974.

Feuerbach's theorem   Feuerbach proved, by calculating their radii and the distances between their centres algebraically, that the nine-point circle touches the incircle and each of the excircles of the triangle. This adds another 4 significant points to the nine-point circle.

The nine-point circle of ABC is also the nine-point circle of the triangles AHB, BHC and CHA, and therefore touches the incircles and excircles of each of these triangles. This adds $3 \times 4 = 12$ more points, giving a grand total of 25. There are more ...

If T is one of the points where the nine-point circle touches the other four circles, and if A, B and C are the mid-points of the sides, then one of the lengths TA, TB, and TC is the sum of the other two.

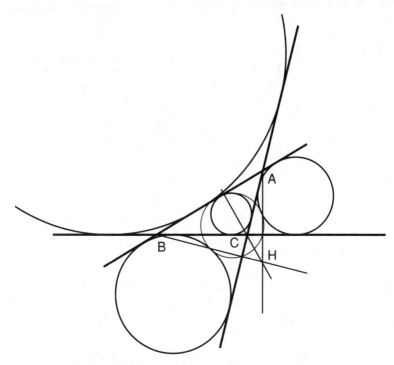

fifty-nine icosahedra   The tetrahedron and cube cannot be stellated because their faces, on being extended, will not again intersect. The octahedron has one stellation, the *stella octangula*, and the dodecahedron has three: the small stellated dodecahedron, the great dodecahedron, and the great stellated dodecahedron.

The icosahedron, in contrast, has no less than 59 stellations, enumerated by M. Bruckner, A. H. Wheeler and H. S. M. Coxeter. If a solid icosahedron is cut by plane cuts from a solid block of wood, 1 + 20 + 30 + 60 + 20 + 60 + 120 + 12 + 30 + 60 + 60 pieces are created. These can be replaced symmetrically to form 32 *reflexible polyhedra* (that is, having planes of symmetry) and 27 solids which come in right-handed and left-handed pairs. These include the original icosahedron, the great icosahedron, and the compounds of five octahedra and ten tetrahedra. The figure shows the third stellation which is a deltahedron.

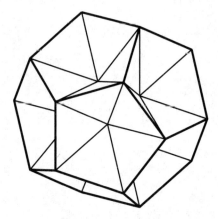

REFERENCE: H. S. M. COXETER, *The Fifty-nine Icosahedra*, Springer-Verlag, Berlin, 1938.

figure-of-eight knot *or* **four-knot**   This is the second simplest knot, with only four crossings, alternately under and over. Join the ends of the knot on the left, and it can be arranged as in the second pattern.

The next sequence shows how the knot above, with one apparent vertical axis of symmetry, is transformed into the third form, which has both a vertical and a horizontal axis of symmetry, and finally into a symmetrical path on the surface of a sphere.

REFERENCE: G. K. FRANCIS, *A Topological Picture Book*, Springer-Verlag, New York, 1987.

five circles theorem   Five circles have been drawn with their centres on the same fixed circle, each of them intersecting the next circle on the fixed circle. By joining the remaining points of intersection, a star pentagon is formed, each of whose vertices lies on one of the five circles.

fixed point theorems   The figure shows a simple example of a fixed point theorem. Two maps have been placed one on top of the other. They show identical regions, but one is larger than the other. The smaller one can be thought of as the result of shrinking the larger one onto a part of itself. This fixed point theorem says that there is one point on the small map which is directly above the same point on the larger map.

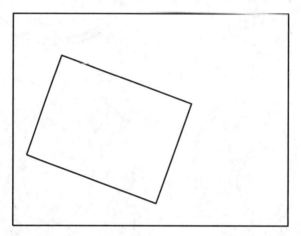

This point (there can be only one) can be found by drawing a third map bearing the same relationship to the smaller map which the smaller map bears to the larger, and then repeating. The sequence of maps tends to a limiting point, which is the point sought.

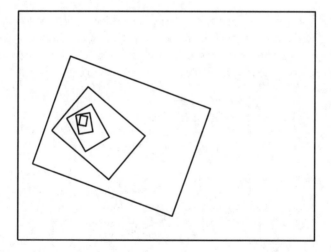

floating bodies in equilibrium   Stanislav Ulam asked whether a sphere is the only solid of uniform density which will float in water in every position. To the simpler problem in two dimensions the answer is 'No!'. A cylinder of density 0·5 with either of these cross-sections will float in water, without tending to rotate, whatever its orientation.

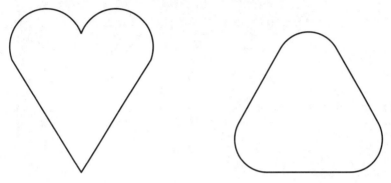

REFERENCE: R. D. MAULDIN (ed.) *The Scottish Book*, Birkhäuser, Boston, 1981.

four colour problem   Any plane map can be coloured with at most four colours, so that any two regions with a common boundary line are different colours.

A map which can be drawn with a continuous line, not taking the pen off the paper, and returning to the starting point, requires only two colours:

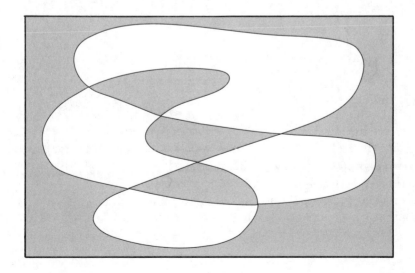

If it does not return to the starting point, it requires three colours:

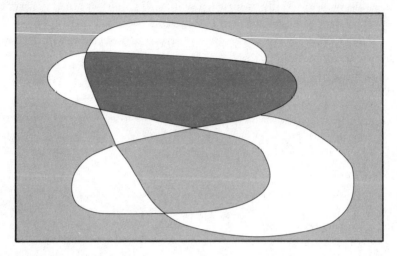

This is the simplest map requiring four different colours:

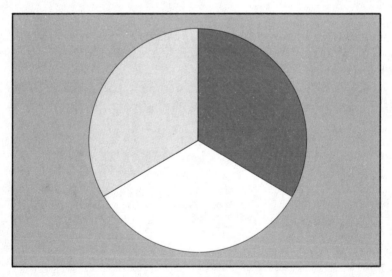

The problem of proving that four colours are sufficient has a long and winding history, including 'what is probably the most famous fallacious proof in the whole of mathematics' announced by Kempe in 1879. For more than a decade it was believed to be sound, until Heawood pointed out the flaw in 1890.

Haken and Appel finally proved in 1976 that four colours are suffi-cient, but only by using a computer program to check several hundred basic maps. This proof has generally been accepted by mathematicians, but only with reluctance, because it is not open to the traditional line-by-line examination that mathematicians have hitherto taken for granted.

Frégier's theorem   Choose any point P on a conic, and make it the vertex of a right angle which rotates about P. Then the lines through the points of intersection, A A, B B, and so on, will all pass through a fixed point X which lies on the normal at P, that is, on the line through P perpendicular to the tangent at P.

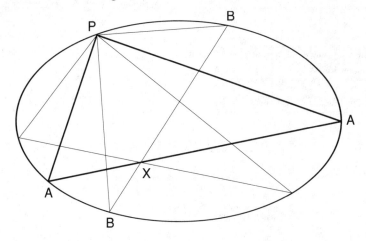

frieze patterns   A frieze consists of a motif repeated *ad infinitum*. If the whole frieze has rotational or reflectional symmetry, then so does the motif of which it is composed.

 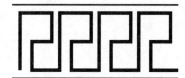

The motif can have no symmetry at all, symmetry about a horizontal or vertical line, or both together, or half-turn symmetry. When motifs are combined in sequence in a frieze, glide reflections, in which the motif moves along the frieze while turning over, produce two more possibilities, making seven types of symmetry in all.

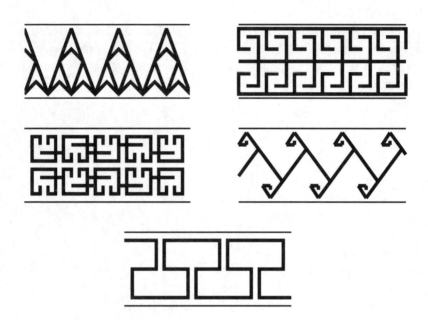

G

Gaussian primes   If p and q are integers, then $p + iq$, where $i = \sqrt{-1}$, is a Gaussian integer. Gaussian integers are either prime, having no proper factors which are also Gaussian integers, or they can be decomposed into Gaussian primes.

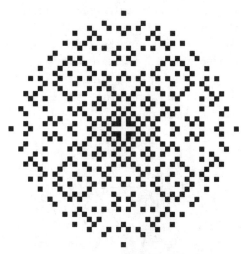

This is the pattern of Gaussian primes whose norms $\sqrt{(p^2 + q^2)}$ are less than 500, drawn on an Argand diagram.
REFERENCE: R. K. GUY, *Unsolved Problems in Number Theory*, Springer-Verlag, New York, 1981.

geodesic dome   Geodesic domes were invented by the engineer-architect Buckminster Fuller. They have the advantage that they can be placed directly on the ground as a complete structure. They also have few limitations of size.

Here is a simple example. Take a regular dodecahedron and its circumsphere. Raise the centre of each face to the circumsphere, and join it by five new equal edges to the vertices of the face. The resulting

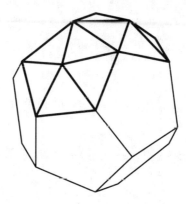

polyhedron has 60 triangular faces, each being isosceles, with edges in the approximate ratio 1 : 1 : 1·115.

Instead of being divided by joining the vertices to the centre of the face, suitably raised, the face may be divided into a larger number of triangular pieces and the vertices of these triangles raised to the circumsphere. In the figure below, each face of an icosahedron has been formed from sixteen smaller almost-equilateral triangles.

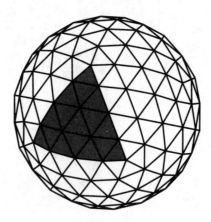

geometrical illusions   If you sketch a figure, it should not always be taken at face value. Figures which are geometrically correct can appear to show something different, and ones which look plausible can in fact be geometrically wrong.

The first figure shows lines which appear to be different in length, but measurement shows that AB and BC are equal.

The second figure has two shaded areas which are equal, although the central disc looks a larger area than the ring. It is easy to prove that they are equal. The circles are drawn in radii increasing by 1 unit. The area of the central disc is $\pi.3^2$ square units and the ring $\pi.5^2 - \pi.4^2 = \pi.3^2$ square units.

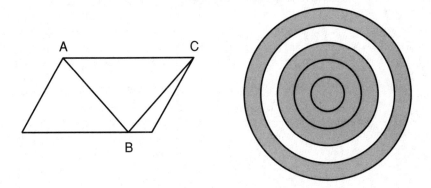

golden ratio, golden section *or* **divine proportion**   If a star pentagon is inscribed in a regular pentagon, the golden ratio naturally appears. The same ratio appears in the dodecahedron and the icosahedron, which Euclid constructed using the division of a line in the 'extreme and mean ratio', as he called it.

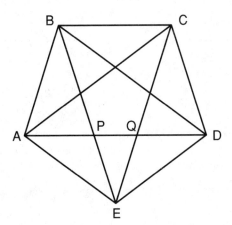

Each of the ratios AQ/QD, AP/PQ and AD/BC is equal to $\frac{1}{2}(1 + \sqrt{5})$, about 1·618. This is usually denoted by the Greek letter ϕ (or sometimes τ).

This ratio has the property that $\phi = 1/(\phi - 1)$ or, expressed in another way, $\phi^2 = \phi + 1$

A 'golden rectangle' whose sides are in this ratio can therefore be dissected into a square and another rectangle of the same shape. The process can be repeated *ad infinitum*.

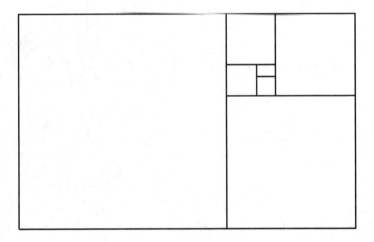

An *equiangular spiral* can be drawn through these vertices. A sequence of circular quadrants is a good approximation to the spiral. The true spiral does not actually touch the sides of the rectangles.

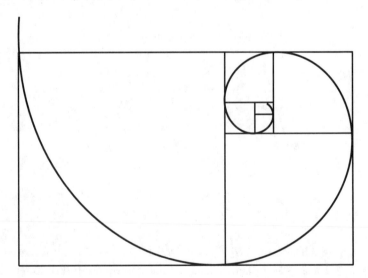

Greek cross tessellation and dissection   The Greek cross tessellates in a very simple manner, which leads naturally to an infinite number of dissections of the cross into a square. Take any four corresponding points of the tessellation, and a dissection of the cross into a square is obtained.

A necessary condition for this simplicity is that the cross is composed of five unit squares where five is the sum of two squares: $5 = 2^2 + 1^2$. This condition is not, however, sufficient. All the other pentominoes (shapes formed by laying five identical squares against each other, complete edge to complete edge) satisfy the same condition, but only some of them will tessellate, leading to similar dissections.

H

hairy ball theorem   This is an example of a fixed point theorem. Imagine that you are combing a tennis ball which is hairy rather than fluffy. You attempt to comb it so that the hairs are all lying flat on the surface and so that they change direction smoothly over the whole surface, but you fail.

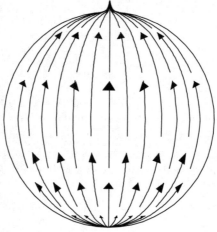

The diagram shows one near-success. You brush upwards from the 'south pole' to the 'north pole', as if brushing along lines of longitude. The entire surface is smoothly combed except at these two points, where a tuft and a hole appear.

Since the Earth is a ball, and the wind at any point has a direction, as if the air were being combed over the Earth's surface, it follows that there is always a cyclone somewhere.

Harborth's tiling   Harborth answered the question: 'Are there sets of tiles which can be used to tile the plane in exactly N ways?'

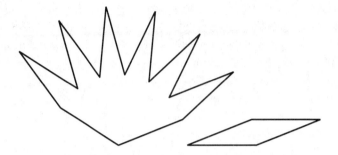

Given these two shapes of tile, a rhombus and 6 rhombuses stuck together, so that 17 of the rhombuses fit round a point, there are exactly 4 ways in which these tiles can tile the plane.

This is one way. Two others are obtained by placing the pair of complex pieces adjacent, or separated by 1 and 4 rhombuses, and the fourth uses just one of the complex pieces.

To construct two tiles that will tile the plane in n ways, use rhombuses of which $6n - 7$ will pack around a point. The complex piece is made by sticking $2n - 2$ rhombuses together, around a point.

REFERENCE: H. HARBORTH, 'Prescribed numbers of tiles and tilings', Mathematical Gazette, No. 418, 1977.

harmonic ratio   Take any two points A and B, and a third point X on the line joining them. Draw any two lines you choose through A and B, to meet at P, and draw PX. Draw AQ and BR to meet on PX. QR cuts AB in another point Y, whose position depends only on the original positions of A, B and X, and not at all on the choice of P, Q and R.

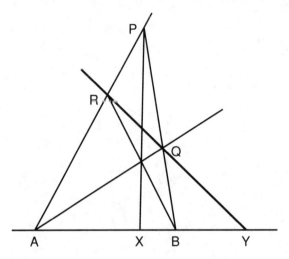

Moreover, the cross-ratio of A, B, X and Y is equal to −1:

$$\frac{AY \cdot XB}{YB \cdot AX} = -1 \quad or \quad \frac{AY}{YB} = -\frac{AX}{XB}$$

The negative sign is because YB is measured in the opposite direction to the other lengths. X and Y are called harmonic conjugates with respect to A and B, and, conversely, A and B are harmonic conjugates with respect to X and Y.

harmonograph   This old Victorian entertainment is revived every few years by some enterprising manufacturer. It requires two pendulums which, in the simplest version, are arranged so that one moves the pen and the other moves the table to which the paper is attached. The combined effect of the two pendulums produces a complicated motion which steadily decays due to the effects of friction. Therefore each path, on each circuit, is a short distance away from the path on the previous circuit, the whole movement tending eventually to a point.

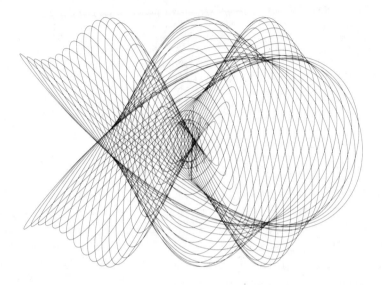

Haüy's construction of polyhedra   The Abbé René-Just Haüy published, in 1784, his 'Essai d'une théorie sur la structure des crystals appliquée à plusieurs genres de substances crystallisées', in which he hypothesized how certain crystals could be built up by regular repetition of a basic unit. These figures show how Haüy ingeniously used small cubic building-blocks to construct the octahedron and rhombic dodecahedron.

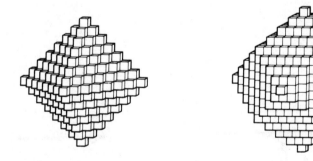

Euclid used the same relationship between the cube and pentagonal dodecahedron in Book XIII of his Elements to construct a regular dodecahedron.

helicoid   When a straight line moves in a screw motion about an axis at right angles to it, it sweeps out a helicoid.

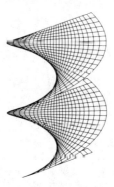

This is a minimal surface. There is an extraordinary connection between the helicoid and the catenoid. The helicoid can be wrapped around the catenoid, as a piece of paper is wrapped around a cylinder. The axis of the helicoid wraps round the circle of smallest cross-section of the catenoid. The second diagram shows how a portion of the helicoid wraps once around the catenoid.

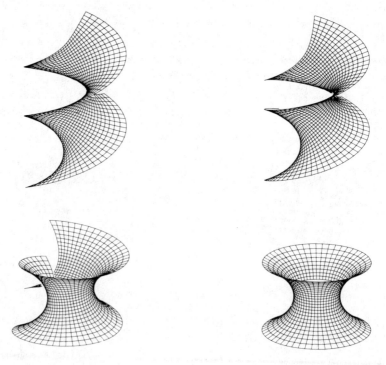

helix   Imagine a circle whose centre moves steadily along a line perpendicular to the plane of the circle. The path of a point which rotates steadily round this circle is a helix. In other words, a helix is the result of a screw motion in a fixed direction.

Depending on the direction of rotation, the helix may be left-handed or right-handed. The figure shows a long cylinder whose axis is helical.

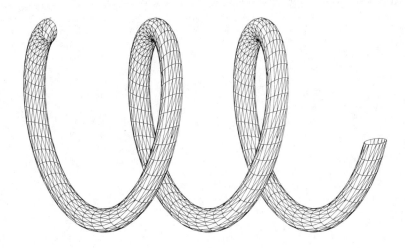

A helix can also be imagined as a curve on the surface of a circular cylinder, which cuts the generators (the straight lines in the surface of the cylinder, parallel to its axis) at a constant angle.

Helices are common in everyday life, because a helix has the useful property that it is transformed into itself by rotating and moving forwards or backwards along its axis. It is therefore the form of the edges of bolts, cylindrical screws and worm gears, as well as a spiral staircase which allows easy movement upwards in a confined space. The curved edges of these shapes are helices, and the curved surfaces are portions of *helicoids* or cylinders.

Hénon attractor   First investigated by the French mathematician Michel Hénon, using an HP-65 programmable calculator, this famous mapping represents the behaviour of many dynamical systems in which no energy is lost, such as asteroids orbiting the Sun.

It is defined by the transformation:

$$x \rightarrow y + 1 - ax^2, \quad y \rightarrow bx$$

Provided the initial point (x, y) is not too far from the origin, after a few repeated applications of the above transformation the point will come to lie within this attractor. With each iteration the point jumps from one curve to another, or to another part of the same curve, in a *chaotic* manner.

Magnifying the map on the computer screen, as in the second figure, shows that each curve is composed of yet finer lines, which in turn are composed of finer lines still ...

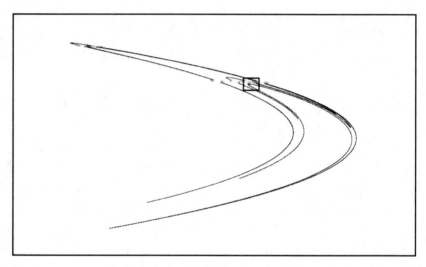

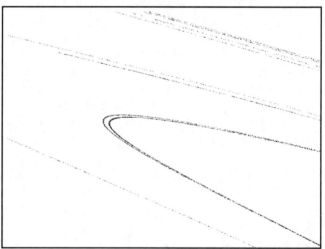

This map shows the Hénon attractor for Hénon's original values, $a = 1·4$ and $b = 0·3$. If all the points on this straight line are transformed by the Hénon process, the line itself is transformed by a process of stretching and folding, rather like the stirring of one liquid into another, into a shape much like the Hénon attractor:

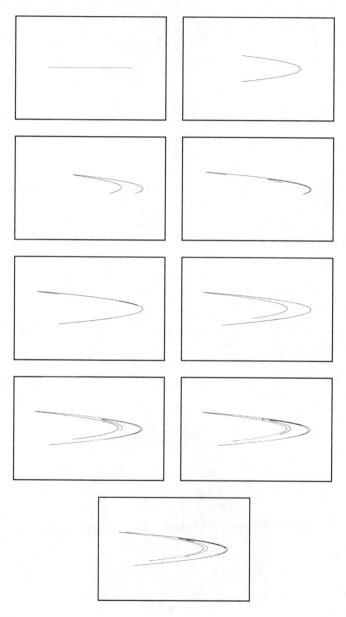

heptahedron   This is a one-sided surface made from four triangles and three quadrilaterals which is topologically equivalent to (that is, it can be continuously deformed into) Steiner's Roman surface, and is much easier to make.

For a regular model, start with a regular octahedron and leave out every other face. The four remaining triangles meet only at their vertices. Now insert the three squares which are cross-sections of the original octahedron through its centre and the edges of its faces. The resulting polyhedron is a closed surface with no boundary, but is only one-sided.

The regular heptahedron can plausibly be classified as a semiregular (Archimedean) polyhedron, because all its faces are regular and all its vertices identical. Unlike the standard Archimedean solids, it is not convex but intersects itself, along the lines where the three squares cross, and even has a triple point at the centre.

Several other Archimedean polyhedra can be constructed in the same way. This model has the square faces, and the hexagonal faces through its centre, of the cuboctahedron.

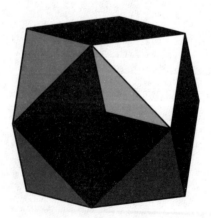

Heron's problem In his *Catoptrica*, Heron of Alexandria assumed that light travels by the shortest path (in terms of distance), and proved that the angles of incidence and reflection at the surface of a mirror are equal.

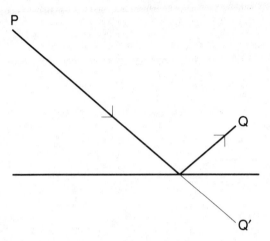

He did so by the same method used today. Reflect the point Q in the mirror. The shortest distance PQ will equal the shortest distance PQ′, which is a straight line. Reflecting Q′ into Q shows that the two angles are equal.

The same principle of reflection solves the problem of finding a point T on a line such that the *difference* between the distances PT and TQ, P and Q being on opposite sides of the line, is as great as possible. T is chosen so that the reflection of Q lies on PT.

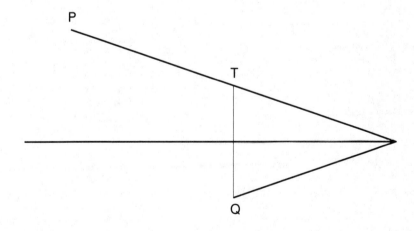

Hilbert's space-filling curve The figures show the first four approximations to the Hilbert curve. The first two show a background of squares which are used to draw the path of the curve. At every stage each square in the previous stage is divided into four smaller squares, and the path is divided so as to pass through the centre of each new square, and replicate on a smaller scale the pattern of the path in the previous stage. In the limit, the result is Hilbert's space-filling curve: a continuous curve passing through every point of the square.

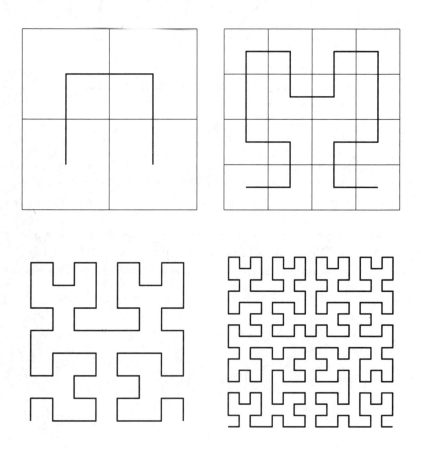

A similar curve which passes through every point of a cube can be constructed in a rather more complicated manner. This is the first stage:

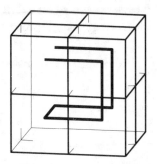

hinged tessellations   Certain tessellations, if they are thought of as being composed of solid pieces hinged at their vertices, and separated by empty space, can be opened out (or closed up) as in these examples. This tessellation of squares and rhombuses is a tessellation of squares, shown near both extremes and at intermediate positions. At two other intermediate positions, each rhombus is equivalent to a pair of equilateral triangles.

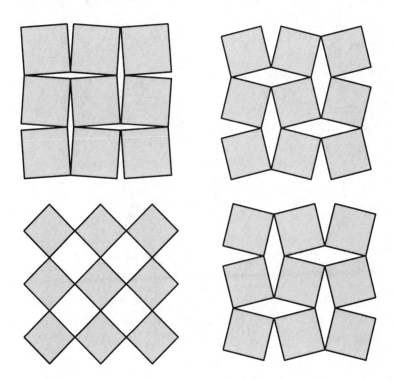

This tessellation of hexagons and triangles hinges in a similar manner. It opens to reveal diamond-shaped spaces which become the squares in a tessellation of hexagons, squares and triangles. If the equilateral triangles continue their rotation, it closes down again to a tessellation of hexagons and triangles, each triangle having rotated through 180°.

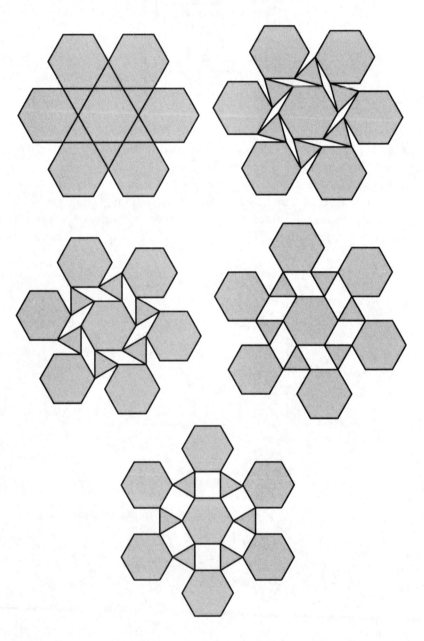

REFERENCE: DAVID WELLS, *Hidden Connections, Double Meanings*, Cambridge University Press, Cambridge, 1988.

Holditch's theorem   Take a smooth, closed convex curve and let a chord of constant length slide around it. Choose a point on the moving chord which divides it into two parts, of lengths p and q. This point will trace out a new closed curve as the chord moves. Then, provided certain simple conditions are satisfied, the area between the two curves will be πpq.

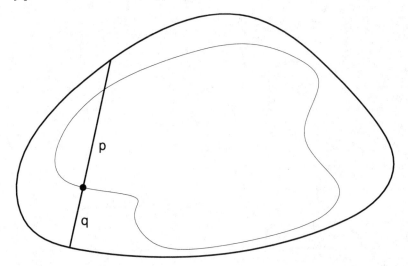

REFERENCE: WILLIAM BENDER, 'The Holditch Curve Tracer', *Mathematics Magazine*, March 1981.

hollow tilings   When a single type of tile inevitably leaves spaces when used in a tessellation by itself, it is tempting to accept the spaces as a feature of the pattern (of course, they could also be considered as a new shape of tile in their own right).

This tessellation was produced by Albrecht Dürer, who, like many Renaissance artists, was fascinated by tilings.

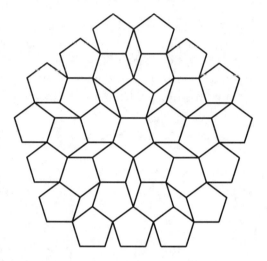

Here is another regular close-packing of pentagons, in which each pentagon touches six others.

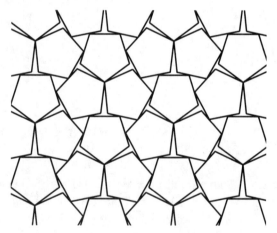

honeycombs   In 1926, Petrie and Coxeter discovered what they called 'regular skew polyhedra' – structures with regular faces and vertices which fill space. In the first case, there are six squares round every vertex, in the next, four hexagons, and in the last, six hexagons round every vertex. Because of their regularity, Coxeter even suggested they should be counted as regular infinite polyhedra; if one includes the three regular plane tessellations, which can also be considered as infinite polyhedra, Coxeter's interpretation brings the number of regular polyhedra to fifteen.

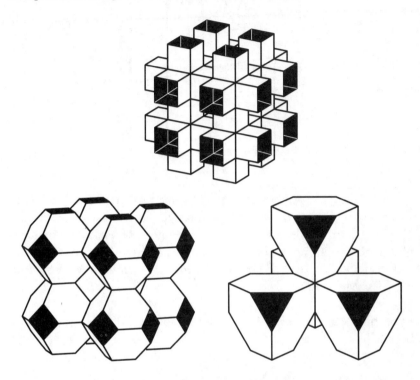

The first two figures are dual, in the sense that the vertices of each are the centres of the faces of the other. The third, like the tetrahedron, is self-dual.

The figure with six squares round each vertex can also be thought of as a standard division of the plane into identical cubes in which every plane is coloured like a chessboard, and all the squares of one colour have been removed. It not only divides space into two congruent halves, but has the extraordinary feature that it is flexible and, if made from individual square faces without some form of stiffening, will collapse down into a plane.

The next figure is a polyhedron whose vertices are all congruent, and which has five squares at each vertex. It lies, as it were, between the square sponge and the square plane tessellation, and has the same relationship to the true sponges that a frieze pattern has to a tessellation.

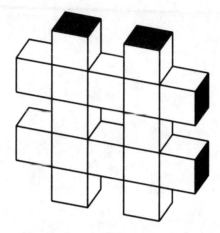

In 1967 J. R. Gott published details of some further, similar, repeating structures, with a slightly different definition. His set included Petrie and Coxeter's, and the previous figure, and three more. One has eight triangles round each vertex, another has ten triangles, and another five pentagons. REFERENCE: J. R. GOTT, 'Pseudopolyhedrons', *American Mathematical Monthly*, May 1967.

hyperbola   The hyperbola is a cross-section of a double cone, cutting both halves of the cone.

A hyperbola has two real asymptotes: two lines which it approaches more and more closely without ever quite reaching them. (An ellipse has two imaginary asymptotes.)

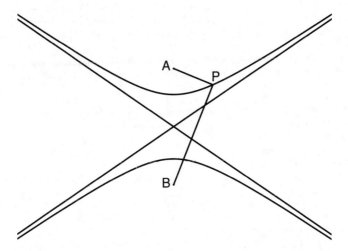

Like an ellipse, a hyperbola has two foci. For any point P on the hyperbola, | PA– PB | is constant.

Like the ellipse and the parabola, the hyperbola may also be defined by its focus–directrix property. Choose a point to be one focus and a straight line to be its directrix. The two branches of the hyperbola are each the path of a point moving so that the ratio of its distance from the focus to its distance from the directrix is greater than one.

The hyperbola can be drawn mechanically by a method similar to, but less simple than, that for the ellipse. Let AX be a rod rotating about A, which will be one focus of the hyperbola. Attach a length of string to the end of the rod and to the other focus, B, and keep it taut by a pencil, shown here at P, held against the rod. As the rod rotates, P traces out one branch of a hyperbola.

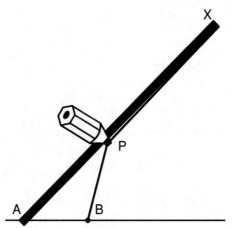

When a ray of light passes through one focus of a hyperbolic mirror, it is reflected as if it had come from the other focus:

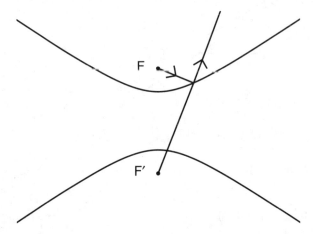

The hyperbola can be constructed as an envelope. Here is one method. Draw a circle and choose a point F, to be one focus of the hyperbola. The diameter through F will be the axis of the hyperbola. Draw any line through F to cut the circle in two points, and draw the perpendicular lines through the cutting points. These lines are tangents to the hyperbola, one to each branch, and by repeating the construction for different lines through F, the hyperbola appears.

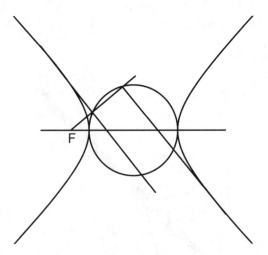

The hyperbola has innumerable other properties. For example, if the tangent to the hyperbola at T cuts the asymptotes at P and Q, and the asymptotes meet at O, then $OP.OQ$ is constant; and $PT = TQ$, as Apollonius showed.

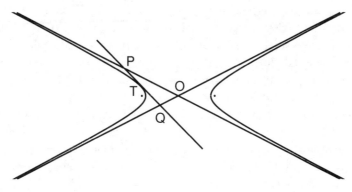

hyperbolic geometry   Euclid in his *Elements* assumed that:

> If a straight line falling on two straight lines makes the interior angles on the same side less than two right angles, the two straight lines, if produced indefinitely, meet on that side on which are the angles less than two right angles.

This is his famous Fifth Postulate, which seems complicated enough to be a theorem, but which neither Euclid nor any of his successors was able to prove.

Bolyai and Lobachevsky independently considered the possibility that it was not provable in principle and that it would make sense to deny it. They each supposed that there were two distinct lines WPX and ZPY, called *limit rays*, through a point P, which do not meet a line AB, such that any line through P within the angle XPY meets AB. Of the lines through P within the angle XPZ, not one of them will meet AB. These

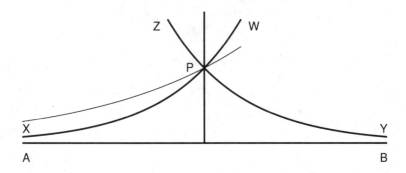

lines they considered as 'parallel' to AB, and so there are an infinite number of lines through P parallel to AB.

This geometry was named 'hyperbolic' in 1871 by Klein. In hyperbolic geometry, the angle sum of a triangle is always less than two right angles. If the triangle is small, then its angles are nearly two right angles.

A triangle is defined by its angles; in hyperbolic geometry there are no such things as similar triangles, because two triangles with the same angles are congruent. The area of a triangle is equal to $K(\pi - \alpha + \beta + \gamma)$, where K is a constant and α, β and γ are the angles of the triangle. The expression $\pi - \alpha + \beta$ is called the *defect* of the triangle. Polygons also have their own defect; two polygons are mutually dissectable if they have the same defect.

A triangle can have three zero angles, all its sides being limit rays of infinite length, and its defect is then a maximum, two right angles. Its area, however, is finite. (Coxeter records that Lewis Carroll could not bring himself to accept this conclusion, and concluded instead that non-Euclidean geometry must be nonsense.)

The circumference of a circle is not proportional to the radius, but increases much faster than the radius, roughly exponentially. However, it is roughly proportional for small radii.

In the limit, as the constant of hyperbolic geometry tends to infinity, hyperbolic space becomes 'flat' and Euclidean. Hence hyperbolic geometry includes Euclidean geometry as a special case. Lobachevsky realized this, and called his new geometry 'pangeometry'.

hyperbolic paraboloid A saddle-shaped quadric surface whose cross-sections in two perpendicular directions are parabolas, and in the third,

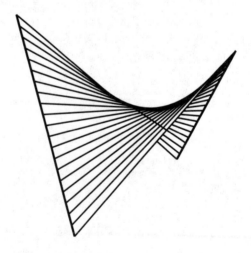

perpendicular direction, hyperbolas. The asymptotes of all these hyperbolas form two planes passing through the common axis of all the parabolas. Like the hyperboloid of revolution, its surface contains two sets of straight lines, called its generators.

A model can be constructed by starting with a skew quadrilateral in three-dimensional space. Two threads joining the mid-points of opposite sides will meet. If the sides are quartered, then pairs of lines joining matching quartering points will also meet each other, and will also meet the lines joining the mid-points. By continuing this process, the surface is generated, each thread being a generator.

The figure shows the skew quadrilateral with one set of generators and one line of the second set, the one at the saddle point.

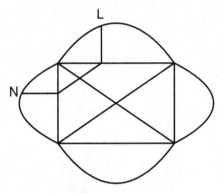

Another method of making a model is described by McCrea. Draw a rectangle, and construct parabolas of equal height on each side. Divide a diagonal into as many equal parts as you choose, and hence find the points of division of the parabolas, such as L and N. Bend two sides up and two down, and join the points, as in the second diagram. This model shows the role of the parabolas more clearly. The two sets of lines are the generators, as before.

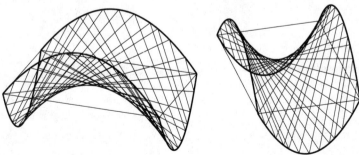

A third approach, mathematically important but impractical, is to start with three skew lines which are parallel to one plane, but not to each other. Through any point on one of these lines, there will be a unique line which cuts both the other lines. The set of all such lines is one set of generators of a hyperbolic paraboloid. The set of lines crossing all the lines of this set, which includes the three original lines, is the second set of generators. REFERENCE: W. H. McCREA, *Analytical Geometry of Three Dimensions*, Oliver and Boyd, Edinburgh, 1947.

hyperboloid of one sheet   Stick a needle through a match, and stick another match on the end of the needle. If the matches are parallel, then when the first match is rotated on its long axis the second will trace out the surface of a cylinder. But if they are not parallel, and if they are not in the same plane, the second will trace out a hyperboloid of revolution of one sheet. (Sir Christopher Wren was the first to realize that the surface of the hyperboloid of revolution of one sheet contains sets of straight lines.) The positions of the second match as it rotates define one set of generators of the surface. No two generators from this set ever meet each other, and no three can be parallel to the same plane.

It may seem intuitively obvious that if the angle of the second match is switched, so that it points backwards instead of forwards, to the same degree, then it will trace out the same surface. It does, and its positions are a second set of generators, each of which intersects every line in the first set (with the exception of one, opposite, line in the first set, to which it is parallel).

Two identical surfaces of this type can be used as the basis for skew-bevel gears, by which a rotating axis can transfer its motion to an axis which is not parallel to it, and does not intersect it. The surfaces are designed so that a generator in one surface aligns with a generator in the second, and the surfaces both roll and slide against each other.

This hyperboloid of revolution naturally has circular cross-sections perpendicular to the axis of rotation. The general hyperboloid of one sheet has elliptical cross-sections.

A practical method of constructing the surface is to take two circles, or ellipses, parallel and on the same axis, with their axes parallel. Divide each ellipse into the same number of parts, by marking equal angles from the centre. If each point in the upper ellipse is joined to the point N steps ahead in the lower ellipse, the lines will form the hyperboloid of one sheet. The second set of generators is added by counting back N steps each time.

If a model is made from rigid wires, rather than threads which require tensioning, then a remarkable feature can be demonstrated. If the ellipses are brought closer together, without rotating relative to each other, so that the wires slide through one set of holes, then the surface remains a hyperboloid, becoming, in the limit, an ellipse and its envelope of tangents.

hypercube *or* **tessaract**   A hypercube is the four-dimensional analogue of a three-dimensional cube. Just as the latter can be obtained by duplicating a square, moving the duplicates apart, and joining corresponding edges, so can a hypercube – by separating duplicate three-dimensional cubes.

On the left below are two congruent cubes, in projection, with corresponding edges joined. On the right, one cube is inside the other; each face

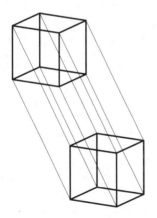 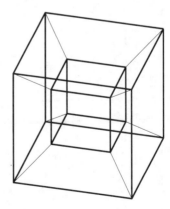

of the outer cube, plus the matching face of the inner cube and the four lines joining them, make up one of the cubical faces of the hypercube. Counting the original two cubes, this is a total of 8 cubical faces, or *cells*. It has 24 plane faces, 32 edges and 16 vertices.

The hypercube has eight main diagonals, joining pairs of opposite vertices. These divide into two sets of four, the diagonals in each set being mutually perpendicular. The dual of the hypercube is the 16-cell.

I

incentres and excentres of a triangle   A unique circle touches the three sides of a triangle internally, and three circles each touch one side externally and the two others internally.

The centres of these circles are the meets of three internal and three external bisectors of the angles of the triangle, which form a larger triangle, with its altitudes.

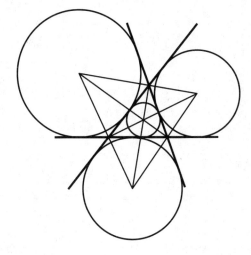

If the radii of these circles are r, r_a, r_b and r_c, then

$$\frac{1}{r} = \frac{1}{r_a} + \frac{1}{r_b} + \frac{1}{r_c}$$

Also, if the radius the of the circumcircle is R, then $r_a + r_b + r_c - r = 4R$, and the area of the triangle is $\sqrt{r_a\, r_b\, r_c\, r}$.

The lines joining the vertices to the points of contact of the inscribed circle meet at Gergonne's point. The lines joining the vertices to the internal points of contact of the escribed circles meet at Nagel's point.

Invert the figure with respect to any of the four circles, and that circle and the sides of the triangle become four equal circles.

The internal bisectors of a triangle define another circle, through the points where they meet the opposite sides. This circle has the property that, of the chords cut off by the sides of the triangle, one is equal to the sum of the other two.

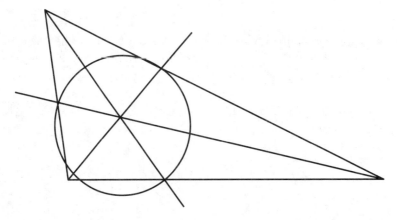

incomparable rectangles   Two rectangles are called incomparable if neither of them will fit inside the other, with their sides parallel. This is equivalent to saying that one of the rectangles is both the longest and the narrowest.

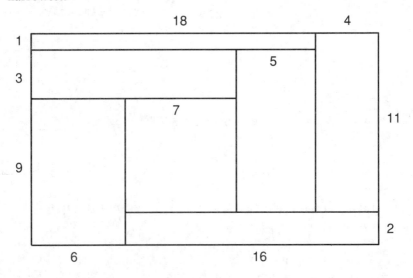

What is the smallest number of mutually incomparable rectangles that will tile a rectangle? At least 7 tiles are necessary, and at most 8. This 13×22 rectangle is the smallest rectangle (whether measured by area or by perimeter) with integer sides that can be incomparably tiled.

interlocking polyominoes   A polyomino is formed by laying a number of identical squares against each other, complete edge to complete edge.

How small can a polyomino be, if a set of duplicates makes a tessellation which is interlocking? The question is ambiguous, because it is not clear whether they should interlock in pairs, or only when they are all in place.

These solutions were found by Bob Newman. The first set interlock individually. The second, a well known pattern, and the third interlock when all the tiles are in place, and also happen to be symmetrical. The fourth pattern involves turning half the tiles over, but uses only 12 units per tile.

REFERENCE: DAVID WELLS, *Recreations in Logic*, Dover, New York, 1979.

intersecting chords of a circle   How simple can an interesting figure be? In this figure, two chords of a circle intersect, and $AX.XC = BX.XD$.

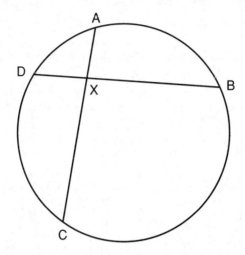

If X is outside the circle, and one of the tangents from X to the circle touches it at T, then $AX.XC = BX.XD = XT^2$.
   Also,

$$\frac{\text{arc } AB + \text{arc } DC}{\text{arc } BC + \text{arc } DA} = \frac{\angle AXB}{\angle CXD}$$

If the chords are perpendicular, then, as Archimedes proved,

$$\text{arc } AB + \text{arc } CD = \text{arc } BC + \text{arc } DA$$

intersecting cylinders   If the axes of three circular cylinders of equal diameter d intersect mutually at right angles, they enclose a solid of 12 curved faces. The volume of this solid is $(2 - \sqrt{2})d^3$.
   If the tangent planes are drawn to all the generators joining vertices where three faces meet, the resulting figure is the rhombic dodecahedron.
   A rather simpler figure is formed when the axes of only two identical cylinders intersect at right angles. Archimedes and the Chinese mathematician Tsu Ch'ung-Chih both knew its volume, which can be found without the use of calculus: $\frac{2}{3}d^3$.
   It is possible for four such cylinders to intersect symmetrically, if their axes have the symmetry of the regular tetrahedron. They form a 24-faced

solid, analogous to the octahedron with inscribed cube, whose volume is:
$\frac{2}{3}(3 + \frac{2}{3} - \frac{4}{2})d^3$.

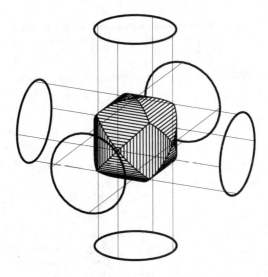

REFERENCE: M. MOORE, 'Symmetrical intersections of right circular cylinders', *Mathematical Gazette*, No. 405, 1974.

inversion Inversion is a transformation of a plane figure into another plane figure, based on a particular circle of inversion whose centre is called the centre of inversion. (In three dimensions, space figures can also be transformed into space figures by using a sphere of inversion.)

If the radius of the circle is k, then the inverse of a point A is the point A' on OA such that $OA.OA' = k^2$.

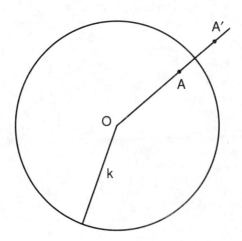

The circle of inversion itself, circles orthogonal to it, and straight lines through its centre are invariant under the transformation. In addition, angles are preserved, and circles and straight lines not through the centre of inversion are all inverted into circles.

The transformation may be used to prove a theorem by transforming it into another one which is either known or obvious. For example, the theorem for Steiner's chain of circles can be proved by inverting the figure into two concentric circles, whereupon the result becomes obvious. Soddy's hexlet can also be inverted. Steiner knew of the process of inversion, but did not reveal its secrets as he stunned his colleagues with a series of surprising and apparently very difficult theorems!

Peaucellier's cell can be used to invert a curve. Many well-known curves are inverses of each other. For example, if a parabola is inverted taking its focus as the centre of inversion, a cardioid results; if it is inverted with respect to its vertex, the result is the cissoid of Diocles.

The next two figures are related by spherical inversion. The pattern of hexagons and triangles below crowds towards two points on the sphere,

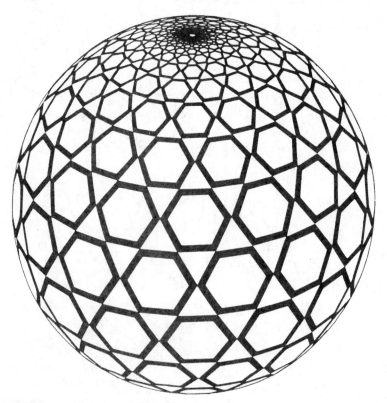

the visible south pole and the north pole. The figure below is the result of inverting the spherical tessellation in the sphere.

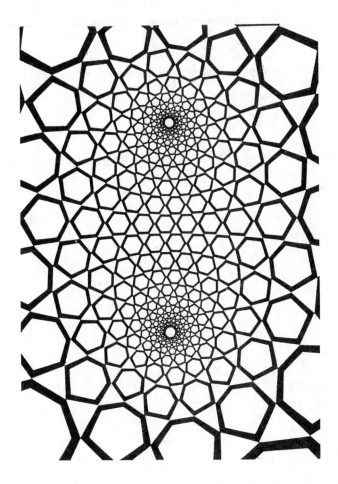

REFERENCE: R. DIXON, *Mathographics*, Basil Blackwell, Oxford, 1987.

Islamic tessellations Islamic artists are well known for their skill and sophistication in using tessellations of all kinds. For example, all seventeen possible *wallpaper patterns* have been found in the Alhambra Palace alone. Many of their patterns involve interlacing.

All such complex patterns can be 'seen' in many different ways. The following pattern can be seen as a pattern of diamonds, each divided into

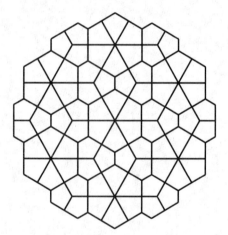

two quadrilaterals and two pentagons, as a pattern of regular hexagons with spokes and truncated equilateral triangles, as a pattern of large hexagons dissected into four small hexagons and seven truncated equilateral triangles … and so on.

isoperimetric problem   The isoperimetric ('equal-perimeter') theorem states that, of all the plane figures with the same perimeter, the circle has the largest area.

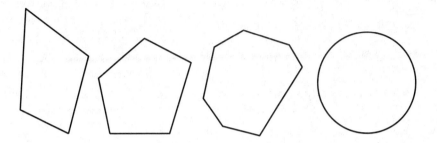

The theorem has a long history. Zenodorus, some time after Archimedes, proved that the area of the circle is larger than that of any polygon having the same perimeter. Pappus also discussed the economy of the bees in constructing their honeycombs, in a famous passage:

> Though God has given to men... the best and most perfect under-standing of wisdom and mathematics, He has allotted a partial share to some of the reasoning creatures as well. To men, as being endowed with reason, He granted that they should do everything in the light of reason and demonstration, but to the other unreasoning creatures He gave only this gift, that each of them should in accordance with a certain natural forethought, obtain so much as is needful for supporting life... That they have contrived [their honeycombs] in accordance with a certain geometrical forethought we may thus infer. They would necessarily think that the figures must all be adjacent one to another and have their sides common... There being, then, three figures capable by themselves of filling up the space around the same point, the triangle, the square and hexagon, the bees in their wisdom chose for their work that which has the most angles, perceiving that it would hold more honey than either of the two others...for the same expen-diture of material in constructing each. But we, claiming a greater share in wisdom than the bees, will investigate a somewhat wider problem, namely that, *of all equilateral and equiangular plane figures having an equal perimeter, that which has the greater number of angles is always greater, and the greatest of them all is the circle having its perimeter equal to them.* (*Mathematical Collection*, Book V)

Steiner finally proved the isoperimetric theorem in several ways in 1841. A related problem is told in the Roman poet Virgil's *Aeneid*: Queen Dido, fleeing her murderous brother, landed on the shores of north Africa, and offered to buy land for herself and her followers from King Jarbas. She was offered as much land as she could enclose with the hide of an ox. According to Virgil, she accepted, cut the ox-hide into a very long thin strip, and enclosed the maximum possible area by using the strip to mark the boundary of a semicircular area against the straight seashore.

REFERENCE: IVOR THOMAS (trans), *Greek Mathematical Works*, Vol. 2, Heinemann, London, 1980.

J

Japanese theorem   Johnson records this Japanese theorem, typical of its period, exhibited in a temple to the glory of the gods and the discoverer, dated about 1800.

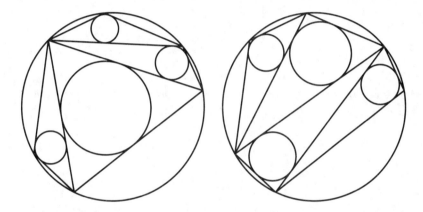

Draw a convex polygon in a circle, and divide it into triangles. Inscribe a circle in each triangle. Then the sum of the radii of all the circles is independent of the vertex from which the triangulation starts. Any triangulation will do: the sum in the second figure is the same as the sum in the first.

REFERENCE: R. A. JOHNSON, *Advanced Euclidean Geometry*, Dover, New York, 1960.

Johnson's theorem   This extremely simple theorem was apparently first discovered by Roger Johnson, as recently as 1916. This suggests a wealth of geometrical properties still lie hidden, waiting to be discovered, two thousand years after Thales found that 'the angle in a semicircle is a right angle'.

If three identical circles pass through a common point, P, then their other three intersections lie on another circle, of the same size.

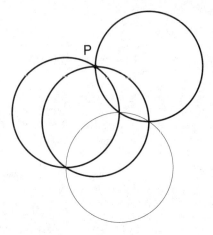

There is a proof as simple as the theorem. Draw the radii, shown in the figure below. These form the skeleton of a cube, because the circles have

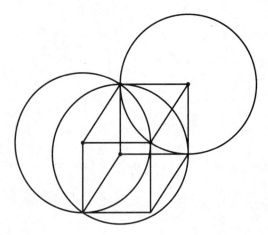

equal radius. Add the missing sides of the cube, and the hidden vertex is the centre of the fourth circle.

Julia set   Choose any complex number, $z = p + iq$, represented by a point (p, q) in the complex plane, and a complex constant k. Calculate $z^2 + k$, take the answer as your new value for z, and calculate $z^2 + k$ for this new value. Repeat, using this third value as the new z ...

This process can be repeated *ad infinitum*. The sequence of values of z, starting with the original value, can be plotted on a graph. What will happen to it? There are three possibilities: it may eventually get further and further from the origin, and disappear towards infinity; it may tend towards a fixed point; or it may end up by jumping around in a region which is called a strange attractor. The strange attractor for a particular point is called its Julia set.

If the original point lies inside the Mandelbrot set, then its Julia set will be a connected set forming a fractal curve, with a fractional dimension. If it lies outside the Mandelbrot set, it will be a set of individual points, called Fatou dust.

The process $z \to z^2 + k$ is the simplest process that will generate this kind of behaviour. However, Julia sets exist for more complex processes. This is the Julia set for the process $z \to \lambda \cos z + k$:

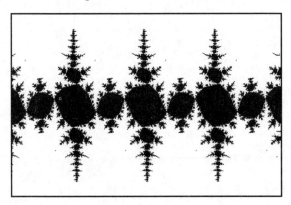

REFERENCE: MICHEL MENDES-FRANCE, 'Nevertheless', *Mathematical Intelligencer*, Vol. 10, No. 4, 1988.

Jung's theorem   The greatest distance between two points in a set is called its diameter. Jung's theorem says that a set whose diameter is 1 unit, or less, can be covered by a circle of diameter $2/\sqrt{3}$ units.

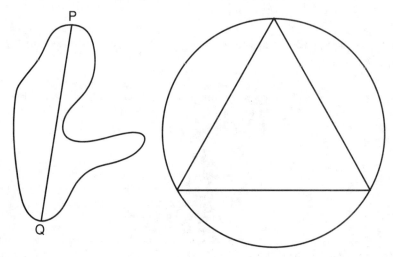

The figure on the left has a diameter PQ. If the equilateral triangle is of side 1 unit, then it is covered by a circle of diameter exactly $2/\sqrt{3}$ units, so that value cannot be improved upon.

K

Kakeya sets *and* **Perron trees**   Kakeya asked in 1917 for the smallest convex region within which a unit line segment could be reversed, that is manoeuvred, so that it rotates completely. Such a region is called a Kakeya set. Kakeya supposed that the answer was an equilateral triangle of unit height. This is correct. What, however, happens if the region does not have to be convex? It was suggested that the answer might be a deltoid of suitable size, in which a unit line could be rotated continuously so that it always touches the deltoid and both its ends lie on the curve, but this conjecture turned out not to be so. The smallest such region has an area which can be made as small as we choose!

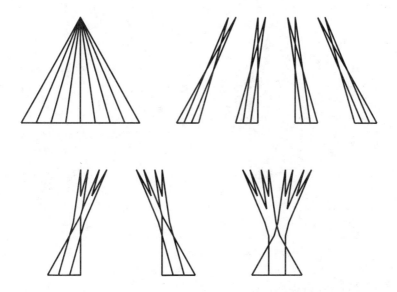

The idea is to halve, and halve, and halve again the base of an equilateral triangle. Adjacent triangles are then slid towards each other so that they overlap a little. The process is repeated with these pairs of triangles, sliding them slightly towards each other.

The result is called a Perron tree. If the base of the triangle is divided sufficiently often, the area of the Perron tree that results can be made as small as we choose. Several Perron trees fitted together provide space for a unit segment to rotate completely.

REFERENCE: K. J. FALCONER, *The Geometry of Fractal Sets*, Cambridge University Press, Cambridge, 1985.

Kepler–Poinsot polyhedra   Pacioli, in his *De Divina Proportione*, for which Leonardo da Vinci is believed to have drawn illustrations, shows an 'elevated' dodecahedron and icosahedron. Pacioli's elevations on the dodecahedron were shallow pentagonal pyramids, and he added regular tetrahedra to the faces of the icosahedron.

Kepler's figures from *Harmonice Mundi* (1619), better known because it contains his third law of planetary motion, illustrated two new polyhedra, which can be considered regular although they are not convex. Their faces are regular star pentagons which intersect each other. They are, on the left, the small stellated dodecahedron, and on the right the great stellated dodecahedron.

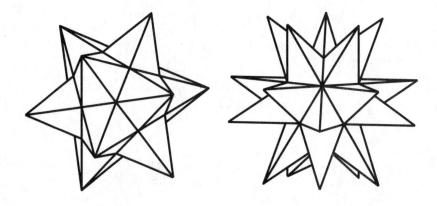

These were rediscovered by Poinsot in 1819, along with two other new non-convex regular solids, the great dodecahedron (left) and the great icosahedron (right).

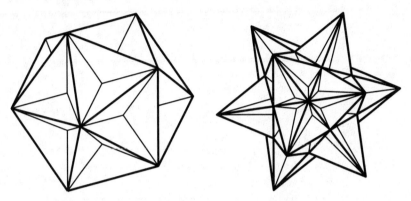

All these solids are three-dimensional analogues of the plane star polygons. The great dodecahedron and the small stellated dodecahedron have bothered some mathematicians because it is not obvious how they fit Euler's relationship, that vertices + faces = edges + 2. Each of them has apparently 12 faces, 12 vertices and 30 edges.

Klein bottle   Take a cylinder and twist one end round so that it passes through its own wall. Join the two ends smoothly, and you have a Klein bottle, named after Felix Klein.

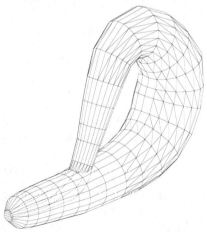

The Klein bottle can be thought of as a rectangle in which one pair of opposite edges have been joined directly, without twisting (CD to C′D′), but the second pair of opposite edges have been joined after a half-twist (AB to B'A').

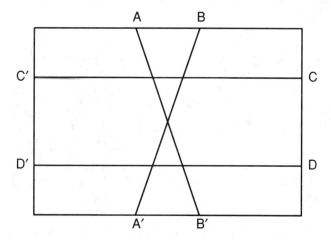

knots   The history of knots is lost in the mists of time. It is quite plausible that human beings used knots before they invented numbers, yet it is only in the last hundred years or so that mathematicians have realized that they are mathematically significant, and have studied them.

This is a bowline knot, a type of knot found a few years ago by archaeologists in a fishing net in Finland and dated by pollen analysts to around 7000 BC. Splice the ends together. The result is the same as if you had started with a sheet bend and spliced its ends: mathematically, the knots are equivalent.

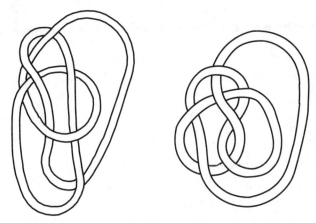

Traditional knots have hundreds of forms and uses, from the severely practical to the intricate and decorative. On the left below is a flat lanyard knot, on the right an 'ocean plait', both reminiscent of the patterns in Celtic strapwork.

knots in sequence　Mathematicians are not interested in whether knots are of practical use (which depends on ease of tying, friction, and so on), so they imagine knots simply as curves in space which do not come apart because the ends have been joined. A curve can be knotted only in three-dimensional space. In four dimensions a curve cannot be knotted, but a surface can be.

A natural way to classify knots is according to their numbers of crossings. These are the *prime knots*, with 7 or fewer crossings. 'Prime' means that the knot cannot be thought of as two knots, tied one after the other on the string.

Notice that any knot can be given an extra but trivial crossing by pinching a small portion and turning it over (either to the right or the left). Such crossings are not counted, and are removed before the knot is classified.

The number of knots with a certain number of crossings increases rapidly, as might be expected. There are 1 each with three (the lowest number possible) and four crossings, 2 with five, 3 with six, 7 with seven, 21 with eight, 49 with nine, and 165 with 10 crossings, if knots which have left-handed and right-handed forms are not distinguished.

Koch's snowflake curve   Take an equilateral triangle, and replace the middle third of each side by two line segments equal in length to the portion removed. Repeat, always replacing the middle third of each straight edge in the same way. Below are shown the first four stages of this 'snowflake curve'. Koch's curve is the limit of this curve as the number of stages tends to infinity.

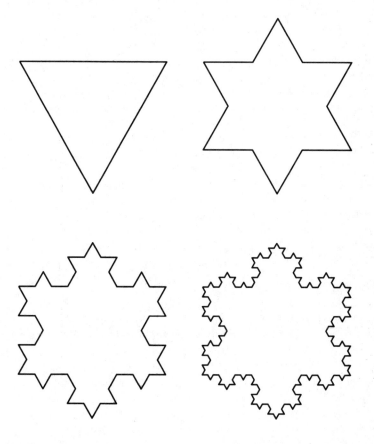

The length of Koch's curve is infinite, but the area it encloses is only $\frac{8}{5}$ of the area of the original triangle. It is a fractal curve, with fractal dimension log 4/log 3, approximately 1·2618 (though ideas of fractals were not around when Koch published his curve in 1904).

The anti-snowflake curve is formed by replacing the middle third of each line by the same two line segments, but pointing inwards. Its area in the limit is $\frac{2}{5}$ that of the original triangle, its length is infinite, and it has an infinite set of double points on the lines joining the centre of the original triangle to its vertices.

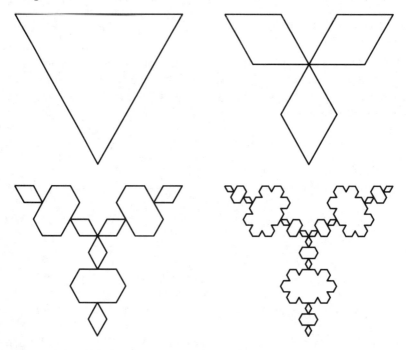

Kürschák's tile   Take a square, and draw equilateral triangles inwards on each of its sides. Find the mid-points of the sides of the square formed by the free vertices of the triangles. These points, together with the meets of the sides of the triangles, are the vertices of a regular dodecagon. The square formed by the free vertices and the inscribed dodecagon forms the basis of Kürschák's tile, shown in the second figure.

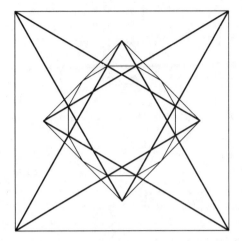

The tile can be used to prove *Kürschák's theorem*: that the area of a regular dodecagon inscribed in a circle of unit radius is 3. (Of the other regular polygons, only the square has a rational area when inscribed in the unit circle.) The entire figure below contains 16 equilateral triangles and 32 isosceles triangles with angles of 15°, 15° and 150°. The 'north' quarter of it contains 4 and 8 of these respectively, which equals the area outside the dodecagon.

The area of the regular dodecagon, 3, gives a rough approximation to π. So does the perimeter of the regular hexagon. I. J. Schoenberg has proved that if a regular n-gon gives a certain approximation to π, by perimeter, then a regular $2n$-gon gives the same approximation by area.

REFERENCE: G. L. ALEXANDERSON and K. SEYDEL, 'Kürschák's tile', *Mathematical Gazette*, No. 421, 1978.

L

Lebesgue's minimal problem   What is the smallest shape that can cover any set of points whose diameter is not greater than 1?

A regular hexagon with side $1/\sqrt{3}$ will do; however, J. Pal proved in 1920 that it is possible to reduce the hexagon slightly, by cutting off the two shaded triangles whose bases touch the inscribed circle. The hexagon with these two triangles removed is called *Pal's universal cover*:

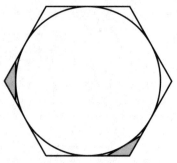

Later Roland Sprague showed that a further small piece could be removed. With centre A draw an arc to touch the opposite edge, meeting the similar arc centre B on the vertical axis of symmetry of the hexagon.

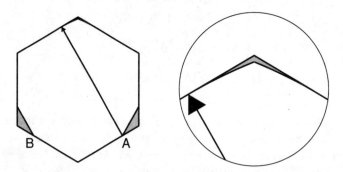

REFERENCE: C. STANLEY OGILVY, *Tomorrow's Math*, 2nd edn, Oxford University Press, New York, 1972.

lemniscate of Bernoulli Named from the Latin *lemniscus,* meaning 'ribbon', by Jakob Bernoulli, in 1694.

To construct the lemniscate as an envelope, start with a rectangular hyperbola, and draw circles whose centres lie on the hyperbola and which go through the centre of the hyperbola. Their envelope is the lemniscate.

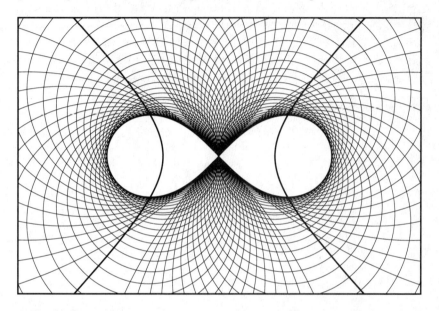

The lemniscate is the inverse of the hyperbola with respect to its centre. Choose a constant, k, and draw a line through O, the centre of a rectangular hyperbola, cutting it at X. Find Y, on OX, such that OX.OY = k^2. The path of Y is the lemniscate.

The polar equation is $r^2 = a^2\cos 2\theta$. It is a special case of *Cassini's ovals.*

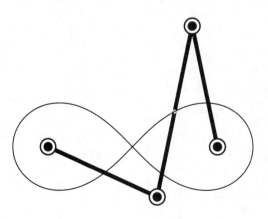

The lemniscate can also be drawn with a very simple linkage. The distance between the two fixed points is equal to the length of the middle rod, and the length of the other rods is $\sqrt{2}$ times this length. The path of the centre of the middle rod is the lemniscate.

limaçon of Pascal   Named after Étienne Pascal, father of Blaise Pascal, though Dürer had already drawn the curve.

Allow a line segment PQ of fixed length to move so that the line, extended if necessary, passes through a fixed point on a circle, and the mid-point of the segment lies on the circle. The limaçon is therefore the conchoid of a circle with respect to a point on it.

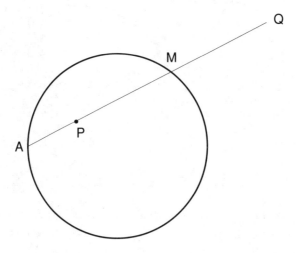

The ends of the segment trace out the limaçon. If the length of the segment is equal to double the diameter of the circle, the limaçon is a cardioid.

It is also generated by a point on a circular wheel rolling round a wheel of the same diameter. It has three forms (like the cycloid), depending on whether the point is on the circumference (generating a cardioid), or inside or outside the wheel.

Its polar equation is: $r = 2a \cos \theta + k$ where a is the radius of the circle and $2k$ is the length of the segment.

To draw the limaçon as an envelope, take a base circle and a fixed point (not on the circle), and draw a circle whose centre lies on the circle and which passes through the fixed point. The envelope of all such circles is the limaçon. (If the fixed point is on the circle, it is the cardioid.)

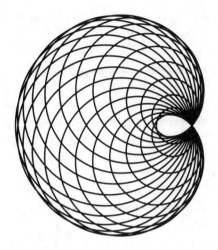

line at infinity   It is often helpful in geometry to think of every line as having a 'point at infinity' and to think of all these points forming the 'line at infinity'. This is thought of as a straight line, rather than a circle.

The 'line at infinity' cannot, of course, be drawn literally. But it can be represented, as in these diagrams, which show first an ellipse which does not meet the line at infinity, then a parabola which touches the line at infinity, then a hyperbola which intersects it.

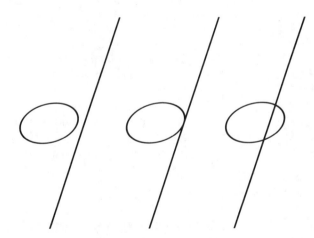

Kepler, that master of analogy, was the first mathematician to think of the conics as forming a continuous sequence, the ellipse becoming, in the extreme, a parabola, which in turn becomes a hyperbola, which is, as it were, an ellipse which disappears to infinity in one direction and then reappears from the opposite direction.

Kepler also introduced the term 'focus' for the special points previously described by Pappus, because a ray of light passing through one focus of an ellipse is reflected through the other focus.

A real circle does not meet the line at infinity in any real points, but does intersect it in a pair of imaginary points, called the *circular points at infinity*. These points are the same for all circles.

The imaginary tangents from the imaginary circular points at infinity to a conic (not a parabola, which touches the line at infinity) form a quadrilateral whose vertices are the four foci of the conic. Two of the foci are real, and two are imaginary.

Lissajous figures *or* **Bowditch curves**　First discussed by Nathaniel Bowditch in 1815, and later by Jules Antoine Lissajous in 1857.

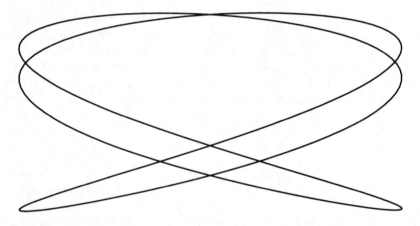

A Lissajous figure is a combination of two harmonic motions in two directions at right angles. If the periods are equal, the curve is an ellipse. If one period is twice the other, the curve is a quartic, including the lemniscate of Bernoulli as a special case.

Its equations can be written in the form

$$x = a \sin(pt + q), \quad y = b \sin t$$

Lorenz attractor　The Lorenz attractor is named after Edward Lorenz, a meteorologist at the Massachusetts Institute of Technology, who discovered it while studying the behaviour of a layer of fluid heated from below, which could be a layer of air in the atmosphere.

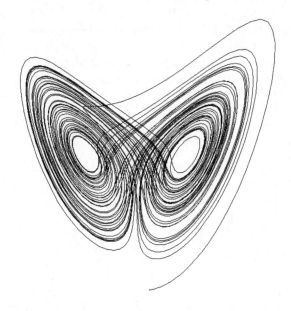

The path of the point representing the behaviour moves in three-dimensional space. It starts at the origin at time zero, swings round one loop, maybe several times, then switches to the other loop, swings around a few times, crosses again, and so on, the switches being apparently unpredictable.

lunes Hippocrates of Chios was described by Aristotle, anticipating one modern caricature of mathematicians, as an excellent geometer but a fool in everyday affairs. More significantly, he is supposed to have been the

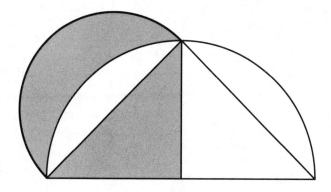

first mathematician to arrange theorems in a logical sequence, an approach followed by Euclid in his *Elements*.

Hippocrates proved that in the figure above, in which half a square is inscribed in a semicircle, with another semicircle on one side of the square, the shaded lune is equal in area to the shaded triangle. (By equating in area a region with a curved boundary and a rectilinear figure, this suggested the possibility of squaring the circle.)

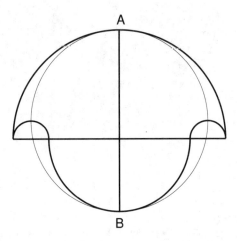

Archimedes proved in his *Book of Lemmas* that this figure composed of semicircles, which he called *salinon* ('salt-cellar'), has an area equal to the circle on AB as diameter.

M

Malfatti's problem In 1803, Malfatti asked for the three largest (in total volume) cylindrical columns that can be cut from a prism of marble. Mathematicians at once assumed that the problem was solved when three circles were found which touched each other and the three sides of the triangular cross-section of the prism. This problem was duly solved by several mathematicians.

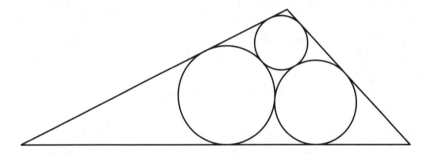

Then, in 1930, it was pointed out that, even in an equilateral triangle, less of the marble is wasted if the columns have cross-sections which are the incircle and two smaller circles, though the increase in the area of the circles is tiny, just over 1% for the equilateral triangle.

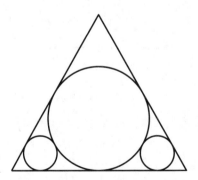

Thirty-five years later (mathematics sometimes moves very slowly) Howard Eves pointed out that if the triangle is long and thin, then this solution is clearly best:

Finally, in 1967 Michael Goldberg proved that the 'original' solution is never best. The maximum area is achieved by one of the alternative arrangements.

REFERENCE: C. STANLEY OGILVY, *Excursions in Geometry*, Oxford University Press, New York, 1969.

maltitudes   The perpendiculars drawn to the sides of a quadrilateral from the mid-points of the opposite sides are called the maltitudes of the quadrilateral. If the quadrilateral is cyclic, then they are concurrent, in the point which is the reflection of the centre O of the circle in the centroid G of the four points.

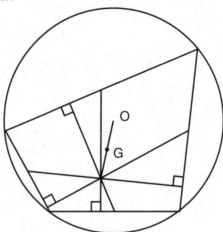

Mandelbrot set   The process $z \to z^2 + k$, which defines the simplest Julia sets, also defines the Mandelbrot set, discovered in 1980 by Benoit Mandelbrot, once a student of Gaston Julia. For each different complex constant k, the origin will either tend to infinity or to a fixed point, or will jump around inside a Julia set. All the values of k for which z does not tend to infinity form the Mandelbrot set. Similar sets exist for different functions. The figure shows the Mandelbrot set (on the left) and the analogous set for the function $z^4 + k$.

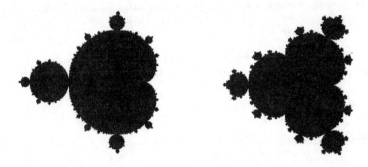

The main body of the Mandelbrot set is a cardioid. This has a large circle attached to it on the left and circle-like regions top and bottom. The cardioid and all these regions have small regions attached to them, and so on, as in Jonathan Swift's verse:

> So, naturalists observe, a flea
> Hath smaller fleas that on him prey;
> And these have smaller fleas to bite 'em,
> And so proceed *ad infinitum*.

The boundary of the Mandelbrot set is fractal. On being blown up on the computer screen (as below), no matter how large the magnification, it is,

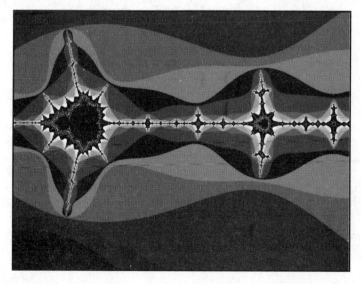

in a sense, similar to itself. In some positions, the whole image of the Mandelbrot appears again. The rate at which z goes to infinity if it is not part of the set can be shown on the screen as a colour or shade of grey. The image shows such an effect, at a magnification of 170, at a point to the left of the circular attached region along the axis of symmetry of the set.

What would happen if, instead of tracing the path of (0, 0) for different values of k, another point were chosen? The result would be just a deformed version of the Mandelbrot set.

Mascheroni constructions   Mohr and Mascheroni discovered the surprising fact that any construction that can be performed with a ruler and compasses can also be done with compasses only. It is also possible to use a ruler alone, without compasses, provided one fixed circle and its centre have already been drawn, or just an arc of a circle (however small) and its centre, or two fixed intersecting circles without their centres.

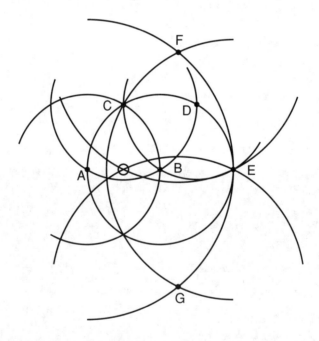

For example, to find the mid-point of AB first construct circles centred at A and B with radii equal to AB. Then draw a circle centred at C with radius AC and a circle centred at D with radius DB. With centre A and radius AE, draw the next circle, followed by the circle centred at E with radius EC. Finally, circles centred at F and G, passing through E, cross at the centre of AB.

REFERENCE: H. STEINHAUS, *Mathematical Snapshots*, 3rd edn, Oxford University Press, 1969.

matchstick constructions   T. R. Dawson, more famous as a chess problemist, discovered that every point that can be constructed with ruler and compasses, and no other points, can be constructed with identical matchsticks, in other words, by the use of identical, movable, straight line segments.

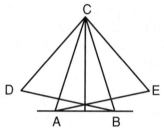

This is Dawson's 7 matchstick construction for bisecting the line AB. As he pointed out, it also serves to bisect the angles ACB and DCE, or any angle less than 120° which is not exactly equal to 60°.

This is how to construct a square. First construct the three equilateral triangles ABC, BCD and BDE. Then let AF be any line within the angle BAC, and construct G and then H. The point F and the crossing of GH and ED define two sides of the required square, BKLM.

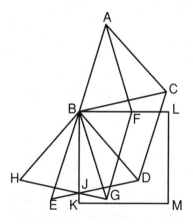

REFERENCE: T. R. DAWSON, '"Match-stick" geometry', *Mathematical Gazette*, No. 254, 1939.

medians of a triangle concur   A median joins a vertex of a triangle to the mid-point of the opposite side. The point where they meet divides each of the medians in the ratio 2 : 1. Curiously, a line from one vertex which bisects a median divides the opposite side in the same ratio.

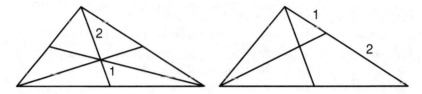

The median point is also the centre of gravity of three equal masses at the vertices, and also the centre of gravity of the whole triangle considered as a sheet of uniform thickness.

Menelaus' theorem   Menelaus of Alexandria proved this theorem in his work on spherical trigonometry.

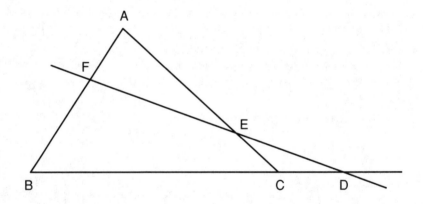

If ABC is any triangle, and DEF is any line cutting all three sides, then

$$\frac{BD}{DC} \cdot \frac{CE}{EA} \cdot \frac{AF}{FB} = -1$$

The ratio is negative because DEF must cut one of the sides externally. In this figure, D cuts BC externally, and DC is measured 'backwards' to C, and so is counted negative. The converse theorem is also true.

Menelaus' theorem generalizes to any polygon.

Miquel point   Four general lines form four triangles, whose circumcircles all concur in their Miquel point, M. This is the focus of the unique parabola which touches the four lines. The centres of the circumcircles also lie on a circle, which passes through the Miquel point.

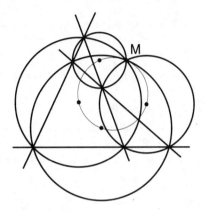

What happens if we start with five general lines? Taking them four at a time produces five Miquel points, which all lie on a circle called the *Miquel circle*. In addition, each set of four lines produces a circumcentre circle, and all five of these circles pass through a common point.

Starting with six lines, each group of five generates a Miquel circle, and, naturally, all these Miquel circles pass through a common point, and so on.

Miquel's theorem   Draw a circle and mark four points on it, A, B, C and D. Draw circles through A and B, B and C, C and D, and D and A. Then the other four intersections of these new circles also lie on a circle.

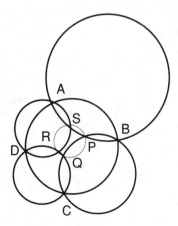

The figure is symmetrical. Although it started with the circle ABCD, it could just as well have started with any of the other circles. This symmetry appears clearly when the points are arranged thus:

A	B	C	D
S	P	Q	R

There are six ways of choosing two letters from the first row. Add the non-matching letters from the second row, and all four points will lie on a circle.

Möbius strip   Take a long, thin rectangle of paper, and join the narrow ends after giving the strip a half-twist. (You could join the longer sides, but this is more difficult.) The result is a Möbius strip, named after August Möbius, who published a description of it in 1865.

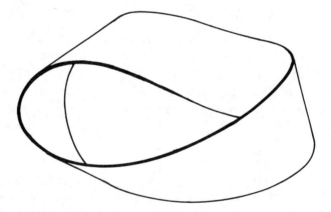

It has one edge, and only one side, and comes in two forms: the right-handed and left-handed, which can be turned over into each other only in four dimensions.

Start at any point on the surface, and draw a line in one direction which does not cross the edge. Keep going, and half-way through your journey you will pass the point you started from, but on the other side of the paper, and after another circuit you will be back to your starting point.

Because it is one-sided, a conveyor belt which is given half a twist, as patented by the Goodrich Tyre Company, will wear evenly on both sides.

Cut a Möbius strip along its centre-line. The result is not two pieces, but one which has *four* half-twists, as if the ends of a long rectangle had been given two full twists before being joined. The edges are now two separate curves, linked to each other but not individually knotted.

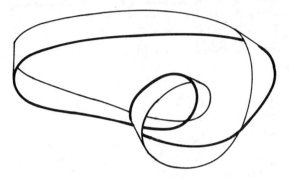

A cylindrical strip of paper can be stretched between two rollers. A Möbius strip stretches round three.

Monge's theorem   The external tangents of three circles x, y and z, taken in pairs, meet in three points A, B and C which lie on a line. If the meets of the internal tangents are included (call them L, M and N, corresponding to the pairs of circles y and z, z and x, x and y), then AMN, BNL and CLM are also straight lines.

L. A. Graham relates that when the engineer John Edson Sweet was first shown this problem,

He paused for a moment and said, 'Yes, that is perfectly self-evident.' Astonished, his friend asked him to explain ... Prof. Sweet, in effect, replied, 'Instead of three circles in a plane, imagine three balls lying on a surface plate. Instead of drawing tangents, imagine a cone wrapped around each pair of balls. The apexes of the three cones will then lie on the surface plate. On top of the balls lay another surface plate. It will rest on the three balls and will be necessarily tangent to each of the three cones, and will contain the apexes of the three cones. Thus the apexes of the three cones will lie in both of the two plates, which is of course a straight line.'

This theorem was first proposed by d'Alembert, and proved by Monge using exactly the same method as Sweet.

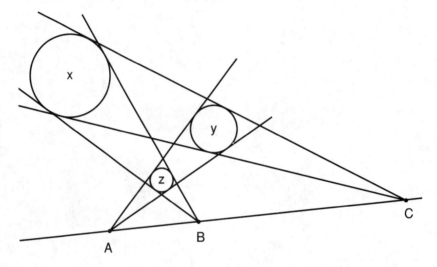

Its analogue in three dimensions says that the apexes of the cones defined by four spheres, taken two at a time, lie in a plane. The cones are drawn so that the spheres are on the same side of the apex.

REFERENCE: I. A. GRAHAM, *Ingenious Problems and Methods*, Dover, New York, 1959.

Morley's triangle   Frank Morley was studying cardioids in 1899 when he came across an extraordinary theorem, which anyone doodling with pencil and paper might have previously spotted.

Take any triangle and trisect the angles. Three of the points where they meet form an equilateral triangle.

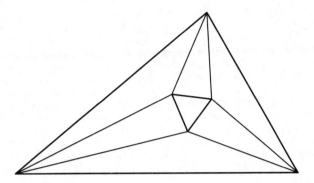

Not surprisingly, if the exterior angles of the triangle are trisected instead, another equilateral triangle is formed. Moreover, the intersections of the other external trisectors and the sides of that triangle form three more equilateral triangles, as in the figure below.

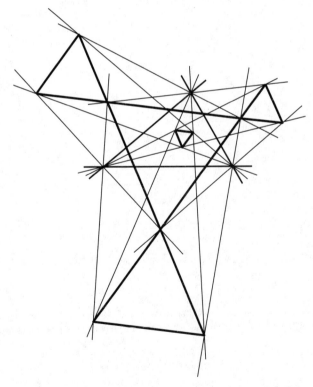

The Morley triangle has the same orientation as the deltoid which is the envelope of all the Simson lines of the triangle.

N

Napoleon's theorem According to legend, Napoleon Bonaparte is supposed to have discovered this theorem. He had some understanding of mathematics, so it is possibly true.

Take any triangle and construct equilateral triangles on its faces. Then the centres of these new triangles form another equilateral triangle.

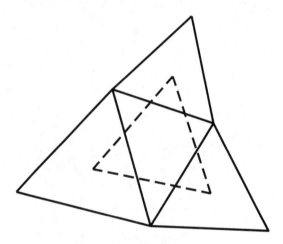

Alternatively, the equilateral triangles can be constructed inwards: their centres still form another equilateral triangle. This triangle has the same centre as the outer triangle, and the difference in their areas is the area of the original triangle.

On the other hand, drawing the centres of one equilateral triangle drawn inwards and two outwards makes a triangle with angles of 30°, 30° and 120°.

The theorem can be proved by embedding the figure in a tessellation, which turns out to have sixfold symmetry.

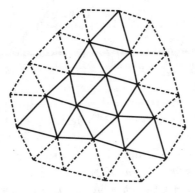

The outer triangles do not have to be equilateral. It is enough that they are similar to each other and attached to the original triangle without changing their orientation, except by slight rotation, as in the figure below. Then the generalized theorem says that corresponding points, one from each triangle, form another triangle of the same shape.

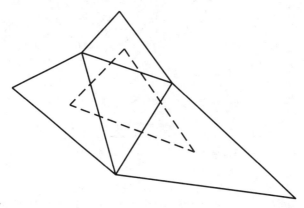

Going back to the original figure, it can be seen to contain many more equilateral triangles:

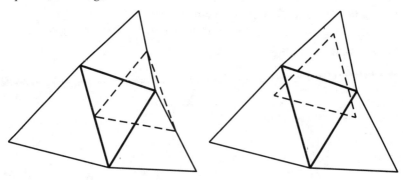

Equilateral triangles constructed on the sides of a triangle are related to the *Fermat point*.

If equilateral triangles are drawn on the sides of a general convex quadrilateral, alternately inwards and outwards, then their vertices form a parallelogram.

nephroid   So named because it resembles a kidney in shape. It is the path of a point on the circumference of a circle of radius a rolling round the outside of a fixed circle of radius $2a$. Alternatively, it is the path of a point on the circumference of a circle of radius $3a$ rolling on a fixed circle of radius $2a$, so that the fixed circle is inside the larger circle.

To draw it as an envelope, draw a base circle, and one of its diameters. Draw a number of circles whose centres lie on the base circle and which touch the chosen diameter. These circles envelope a nephroid.

It is also the envelope of the diameter of one circle which rolls round the outside of another, equal, fixed circle.

The evolute of a nephroid is another nephroid, with the same centre, but half the size and rotated through 180°.

nine-point circle   In any triangle, the mid-points of the sides, the feet of the altitudes, and the mid-points of the lines joining the vertices to the orthocentre, all lie on a circle.

Brianchon and Poncelet published the theorem in 1821, though an otherwise unknown Englishman, Benjamin Bevan, proposed a problem in 1804 which is practically equivalent.

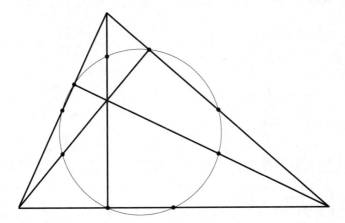

The nine-point circle is half the size of the circumcircle of the triangle, and its centre is half-way between the circumcentre and the orthocentre.

The nine-point circle actually contains many more than nine points: see Feuerbach's theorem.

nine–three configurations   There are three essentially different nine–three (9_3) configurations. That is, there are three ways of arranging 9 lines and 9 points so that there are 3 lines through every point and 3 points on every line.

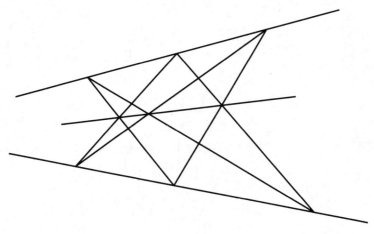

The second of these configurations can be thought of as a triangle which is inscribed in another triangle, which is inscribed in a third triangle, which is inscribed in the first triangle. The first of them is also the figure for *Pappus' theorem.*

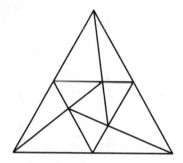 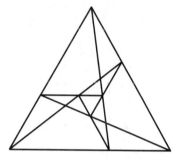

non-alternating knots   A prime knot with fewer than eight crossings must be alternating. That is, if you trace the path of the string as it crosses itself, it goes alternately under-over-under-over. (As when classifying knots by their crossings, trivial cross-overs formed by pinching a piece of string and turning it over are not counted.) The smallest non-alternating knots are these three with eight crossings.

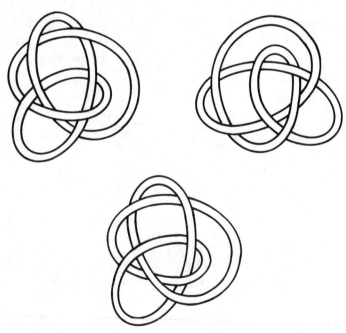

non-rigid polyhedra   Cauchy proved in 1813 that a convex polyhedron made from rigid faces hinged along their edges is rigid. However, if it is not convex there are various alternative possibilities: it may be rigid, or 'shaky' (infinitesimally movable), have two or more stable forms, or be continuously movable, like a linkage.

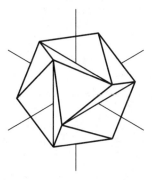

Consider a pair of adjacent faces of an icosahedron. Their edges form a skew quadrilateral, which is also the edge of another pair of triangles with a common edge.

Take a regular icosahedron and replace each of six pairs of faces with an edge in common (matching in orientation the faces of a cube) by a pair of equilateral triangles. The result is *Jessen's orthogonal icosahedron.* It is shaky: it can be infinitesimally deformed, slightly changing the angles along the long edges, without straining on the faces.

Take ten equilateral triangles and make two pentagonal pyramids, joined face to face, except that a gap is left rather than joining the last two pairs. Take two of these incomplete pyramids, and join them at right angles, so that when one is squeezed to reduce its height and widen the

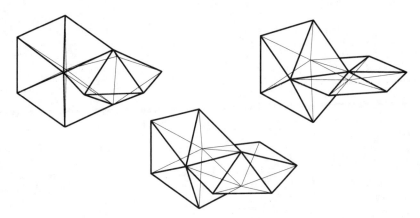

gap, the other becomes fatter. Calculation shows that there are three stable positions for this polyhedron, which was first constructed by Michael Goldberg.

REFERENCE: MICHAEL GOLDBERG, 'Unstable polyhedral structures', *Mathematics Magazine*, May 1978.

O

octahedron   If the edges of a regular octahedron are divided in the golden section (that is in the ratio $1 : \frac{1}{2}(1 + \sqrt{5})$) so that the points of division for any face of the octahedron form an equilateral triangle, then the 12 points of division are the vertices of a regular icosahedron.

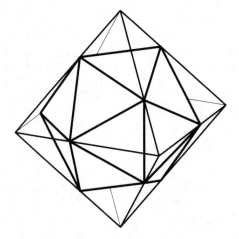

There are two ways in which the edges can be divided internally in the golden ratio, and two more ways in which they can be divided externally, producing a total of four icosahedrons. For external division, the points of division of the edges of one face are next-but-one vertices of the icosahedron.

one-sided surfaces   A one-sided surface may have only one edge, which may be knotted or unknotted. If it has two edges there are more possibilities: each may be knotted or unknotted, and the edges may be linked or unlinked. There are no fewer than eight surfaces with these combinations of possibilities. The first figure is the Möbius strip.

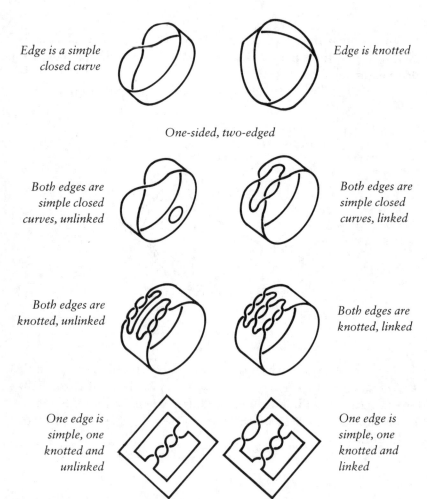

One-sided, one-edged

Edge is a simple closed curve   *Edge is knotted*

One-sided, two-edged

Both edges are simple closed curves, unlinked   *Both edges are simple closed curves, linked*

Both edges are knotted, unlinked   *Both edges are knotted, linked*

One edge is simple, one knotted and unlinked   *One edge is simple, one knotted and linked*

orthocentric points Proclus first recorded that the altitudes of a triangle concur, in the orthocentre (usually denoted by H) of the triangle. The three vertices of the original triangle and the orthocentre possess a beautiful symmetry, as Carnot first noted: any one is the orthocentre of the triangle formed by the other three. They form a set of orthocentric points.

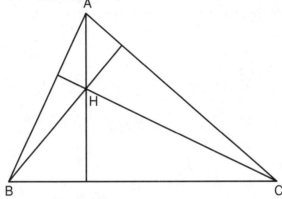

The four triangles formed by four orthocentric points have the same nine-point circle, which their 16 inscribed and escribed circles all touch. The reflection of H in BC lies on the circumcircle of triangle ABC, and so on, and the same four triangles also have circumcircles of the same size.

The centres of the circumcircles of the four triangles form a figure congruent to the original four points, being their reflection in the nine-point centre. The centroids of the four triangles form an orthocentric set, similar to the original set but one-third the size.

Dwight Paine has presented the best-known properties of the triangle in verse. These are his lines on the altitudes, which may tempt readers to look up the rest:

> Although the altitudes are three,
> Remarks my daughter Rachel,
> One point'll lie on all of them:
> The orthocentre H'll.

REFERENCE: DWIGHT PAINE, 'Triangle rhyme', *Mathematics Magazine*, September 1983.

orthogonal surfaces   In two dimensions, two sets, or families, of curves may have the property that each curve of one family intersects every member of the other family orthogonally. In three dimensions, up to three families of surfaces may have the analogous property. Any pair of surfaces, not from the same family, intersect each other orthogonally along a curve.

The simplest example of three families of orthogonal surfaces is three sets of parallel planes, for example the three sets of planes parallel to the faces of a cube.

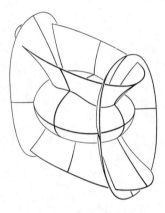

The more complicated example shown here is a three-dimensional ana-logue of the families of confocal conics. It shows one member from each of a family of ellipsoids, a family of hyperboloids of one sheet, and a family of hyperboloids of two sheets. One surface of each family passes through any point in space. Space is therefore divided into curvilinear boxes, whose vertex angles are all right angles, as in an ordinary rectangular box, but whose faces are not even flat, let alone rectangular.

P

packing circles rigidly Arrange identical circles to form an infinite hexagonal tessellation, with gaps, and remove every third circle. Then replace every remaining circle by three small circles, touching each other and the adjacent circles. The figure shows the result of replacing some of the original circles by triples of smaller ones.

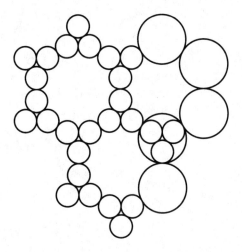

The resulting pattern of circles is rigid, in the sense that each circle is securely held by three adjacent ones. The completed pattern of small circles covers only $(7\sqrt{3} - 12)\pi$, or approximately 0·393 of the plane, probably the lowest possible density for such a rigid packing of circles.

REFERENCE: H. MESCHKOWSKI, *Unsolved and Unsolvable Problems in Geometry*, Oliver & Boyd, London, 1966.

pantograph The pantograph exploits the properties of similar triangles in order to produce an enlarged or reduced copy of a figure. The point O in the next figure is fixed; AB is parallel to CD, and OA to DE. If D traces out one figure, then the point B traces out another similar figure, enlarged in the ratio OB : OD.

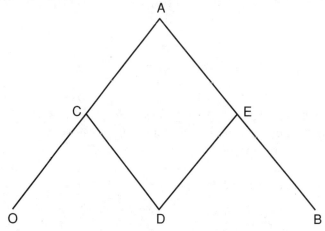

Pappus' theorem   Take any two lines and three points on each, and cross-join the points, as in the figure. The meets of the cross-joins lie on a straight line. This is a special case of Pascal's theorem, because the original pair of straight lines can be thought of as a special case of a conic.

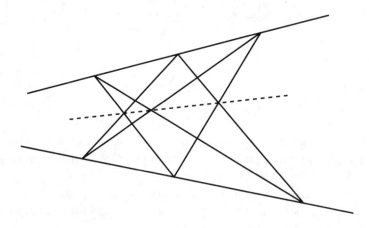

Because the theorem involves only points, straight lines, meets and joins, it has a dual, in which points are swopped for lines, and lines for points. The following figure is a special case of the dual theorem, with the three lines through each of the two points being two sets of three parallel lines which meet in two points at infinity. Joining the points shown in the figure gives three lines which meet in a point.

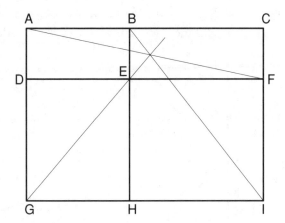

In this figure a rectangle is divided into four smaller rectangles. The thin lines meet at a point. So do the lines DB, GC and HF, which illustrates the fact that in the original Pappus figures the points on each line can be considered in different orders.

Pappus' theorem is equivalent to the following theorem about points and lines in three dimensions. Take three lines in space, a, b and c, which do not meet. It is possible to find an infinite number of other lines which meet all three of them. Choose three such lines, and call them p, q and r. Then the equivalent theorem says that if d is a fourth line meeting p, q and r, and s is a fourth line meeting a, b and c, then s and d also meet.

parabola   The Greeks considered the parabola to be a section of a right-angled cone, parallel to a line through its vertex. In fact, it is the cross-section of any cone by a plane parallel to a straight line through the vertex of the cone.

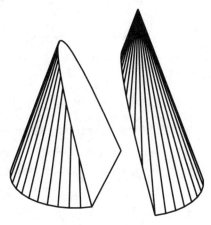

It is also the path of a point moving so that its distance from a fixed point, called its focus, equals its distance from a fixed line, called the directrix:

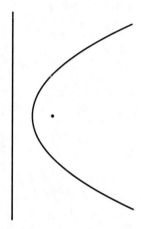

A ray of light passing through the focus will be reflected from a parabolic mirror parallel to the axis. Therefore headlamps use approximately parabolic mirrors with the source of light at the focus.

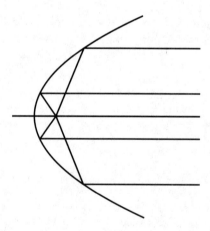

If we assume that gravity acts directly vertically, rather than towards the centre of the Earth, and we ignore air resistance, then we may consider the path of a projectile to be a parabola. If shells are fired in all directions from one point, with the same constant initial velocity, then the envelope of their paths is another parabola, or, in three dimensions, a paraboloid of revolution.

The surface of a slowly rotating liquid in a circular bowl is a paraboloid of revolution. Any vertical cross-section of this surface will be a parabola. A parabola is also the shape of the main cables of a uniformly loaded suspension bridge, if the weight of the cables and supports is ignored.

The parabola may be constructed as an envelope. For example, draw two lines and mark off equal lengths from their intersection, and number them in opposite directions, as in the figure. Straight lines drawn between points with the same number will envelope a parabola. The two original lines are tangents. To draw the more of the parabola, mark further points past the intersections of the original lines as shown in the right-hand figure.

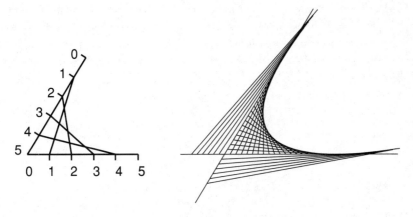

This construction works because of this simple and elegant property of the parabola. Draw three tangents, as in the figure. Then SP/PA = QO/OP = BQ/QS.

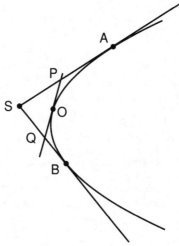

A similar method is used to draw the tangents from a point to a parabola. Join the point P to the focus F and construct the circle whose diameter is PF. If the tangent at the vertex of the parabola cuts the circle at A and B, then PA and PB are the two tangents.

A parabolic envelope is formed when a right-angled set-square is moved so that the hypotenuse passes through a fixed point, which will be the focus, and the opposite vertex lies on a fixed line, the directrix.

Pascal configuration   Six points on a conic can be chosen in order as the vertices of a hexagon in 60 different ways. Each choice generates a *Pascal line*, according to Pascal's theorem. These 60 lines form a complex configuration. The figure shows the Pascal configuration for six points arranged as a regular hexagon on a circle, so that the complexity is reduced. This results in a number of cases of degenerate lines. The three bold lines each consist of four degenerate Pascal lines. A further six lines are the line at infinity, so only 45 lines are visible.

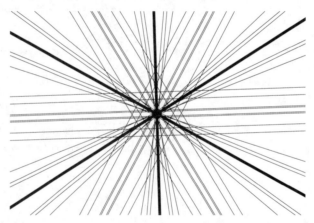

The next figure shows the centre of the first one magnified, to show some of the ways the Pascal lines meet. They pass three at a time through the 20 Steiner points, and also three at a time through the 60 Kirkman points. Each Steiner point lies, together with three of the Kirkman points, on a Cayley line, of which there are 20. The Steiner points also lie, four at a time, on the 15 Plücker lines, and the 20 Cayley lines pass four at a time through 15 points, called the Salmon points.

There is a symmetrical relationship here between the 60 Pascal lines and 60 Kirkman points, the 20 Cayley lines and 20 Steiner points, and the 15 Plücker lines and 15 Salmon points.

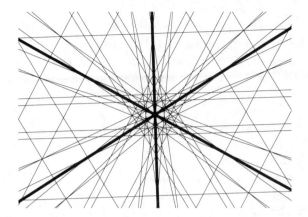

Pascal's theorem   Blaise Pascal discovered his famous theorem at the age of 16, in 1640, and published it as a small pamphlet entitled *Essai pour les coniques*. The theorem states that if a hexagon is inscribed in a conic, then the three points in which pairs of opposite sides meet will lie on a straight line. If the points of the hexagon are labelled in order as ABCDEF, then AB and DE are opposite sides intersecting in X and so on. The line XYZ is then the Pascal line.

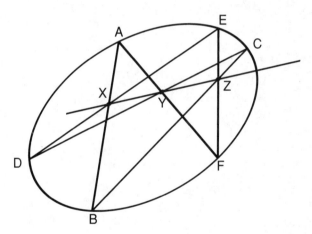

For a zigzag-inscribed hexagon, the points of meeting are inside the conic and the figure looks much like the figure for Pappus' theorem. Indeed, Pappus' theorem is a special case of Pascal's theorem in which the conic degenerates into a pair of straight lines. If the hexagon is drawn in a more normal manner, then the three collinear points lie outside the conic.

Pascal's triangle Write down Pascal's triangle (for more details, see the *Penguin Dictionary of Curious and Interesting Numbers*) on hexagonal paper. Shade in the hexagons containing odd numbers. The pattern on the left results (as generated by a computer using a dot instead of a hexagon).

Shading multiples of numbers other than 2 produces different patterns. At centre and right are the patterns for multiples of 7 and 8.

Pascal's triangle as a pattern of odd and even numbers can be thought of as a cellular automaton. Given a row of 0s and 1s in an infinite strip of cells, the next row of cells in filled by the rule that adjacent 1s or adjacent 0s have a 0 beneath them, and the pairs 01 or 10 have a 1 beneath them. Any pattern of 1s and 0s will do to start with.

```
1 0 0 0 1 0 1 0 1 1 0 1 1 0 1 1 1 1
 1 0 0 1 1 1 1 1 0 1 1 0 1 1 0 0 0
  1 0 1 0 0 0 0 1 1 0 1 1 0 1 0 0
   1 1 1 0 0 0 1 0 1 1 0 1 1 1 0
    0 0 1 0 0 1 1 1 0 1 1 0 0 1
     0 1 1 0 1 0 0 1 1 0 1 0 1
      1 0 1 1 1 0 1 0 1 1 1 1
       1 1 0 0 1 1 1 1 0 0 0
        0 1 0 1 0 0 0 1 0 0
         1 1 1 1 0 0 1 1 0
          0 0 0 1 0 1 0 1
           0 0 1 1 1 1 1
            0 1 0 0 0 0
             1 1 0 0 0
              0 1 0 0
               1 1 0
                0 1
                 1
```

It is easy to change the rule. In the following pattern, an odd number of 1s in three adjacent cells generates a 1 in the cell below the middle cell of the three; an even number of 1s generates a 0.

```
1 1 1 0 1 0 0 1 1 0 0 1 1 1 1 0 0 0
 1 0 0 1 1 1 0 0 1 1 0 1 1 0 1 0
  1 1 0 1 0 1 1 0 0 0 0 0 0 1
   0 0 1 0 0 0 1 0 0 0 0 1
    1 1 1 0 1 1 1 0 0 1
     1 0 0 0 1 0 1 1
      1 0 1 1 0 0
       0 0 0 1
        0 1
```

pencils of conics   There is a unique conic through 5 points, or touching 5 lines, and there is an infinite family of conics touching four lines.

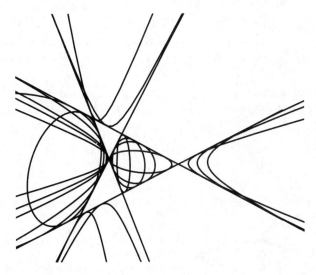

Of the eleven regions into which the four lines divide the plane, only five contain a conic touching all four lines. Parabolas occur only in the left-hand region, which also contains ellipses and portions of hyperbolas. The only closed quadrilateral region contains ellipses only.

Penrose tilings   Roger Penrose, a world famous specialist in relativity and quantum theory, is also an enthusiast for mathematical recreations. He and his father invented the Penrose staircase, which ascends for ever and ever. Penrose discovered these two tiles, felicitously named 'darts' and 'kites' by John Conway.

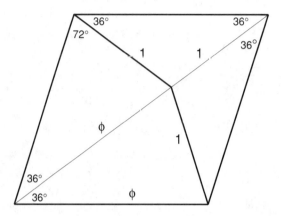

They are constructed from a rhombus. The length ϕ is the golden ratio, $\frac{1}{2}(1 + \sqrt{5})$ or 1·618... . To make a Penrose tiling, the vertices are labelled H for heads or T for tails. The tiles are then assembled so that no two vertices with the same label are ever adjacent. Alternatively, the edges can be nicked (or slightly distorted) in order to ensure this method of assembly, or curves can be drawn on each tile, as in the figure below, so that when the tiles are correctly assembled the curves on one tile joins up with the curves on adjacent tiles.

The two shapes can be used to tile the plane in an infinite number of ways which are not periodic. In other words, if you make a transparency of one of these non-periodic tessellations, there is no way you can move the transparency, without rotating it, so that all the lines once again fit the tessellation.

Some Penrose tilings have rotational symmetry; most do not. They all require more kites than darts, roughly in the ratio ϕ : 1. In an infinite tiling, this ratio is exact.

If the curves are drawn on the tiles, then if a curve closes, it has pentagonal symmetry, and so does the region inside the curve.

Any finite region of a pattern can be tiled in only one way, so the outline of a region defines the tiling within it.

Any finite region of one tiling appears not just once, but an infinite number of times in every other infinite Penrose tiling. This has the remarkable consequence that if you descend onto a Penrose tiling and start to explore it, you can never decide which Penrose tiling you are actually on! Moreover, should you decide to explore this new Penrose world in search of particular region with which you are already familiar, you will certainly find it, lying within a distance of at most twice the diameter of the region.

pentagon tessellations   The regular pentagon will not tessellate. Less regular pentagons may, as in the *Cairo tessellation*. How many different types of tessellation are possible with irregular pentagons?

K. Reinhardt found five distinct types of tile in 1918. Richard Kershner, in 1967, found three more which had previously been missed, and believed that his list was complete. However, in 1975 Richard E. James III discovered the beautiful new tiling shown in the next figure.

In this tiling $A = 90°$, $C + D = 270°$, $2D + E = 2C + B = 360°$, and $AE = AB = BC + DE$.

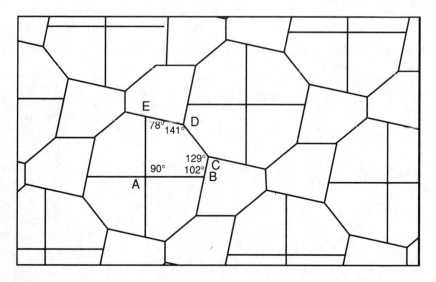

Then, in 1976 Marjorie Rice, according to Martin Gardner, 'a San Diego housewife with no mathematical training beyond the minimum required in high school', found a tenth type, illustrated here, quickly followed by three more.

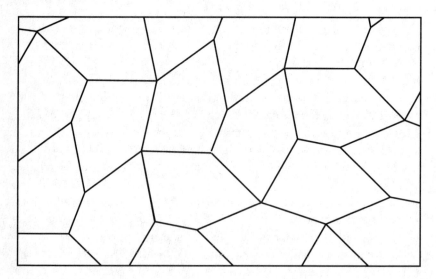

In 1985, Rolf Stein found this fourteenth type. It is not known whether the list is complete.

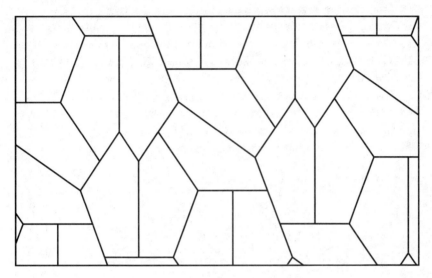

pentatope *or* **simplex**   This is the analogue in four dimensions of the tetrahedron in three dimensions and the triangle in two.

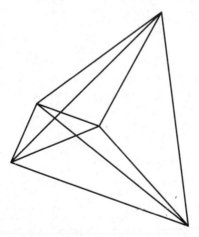

It has 5 three-dimensional faces or cells, each in the shape of a regular tetrahedron, 10 plane faces, 10 edges and 5 vertices, and is its own dual.

Its net in three-dimensional space is, as would be expected by analogy with the plane net of a regular tetrahedron, a regular tetrahedron with a regular tetrahedron stuck onto each face.

If it is regarded as the result of joining two loops of five plane faces each, then each loop of five faces forms a *Möbius strip*.

Truncating a regular tetrahedron, by symmetrically slicing off each vertex, produces a new equilateral triangle at each vertex and changes each face into a hexagon. If the slices are made through the mid-points of the edges, the original faces also become equilateral triangles and the truncated solid is a regular octahedron.

Truncating a pentatope produces new tetrahedra at each of the original vertices and changes its 3-dimensional faces from tetrahedra to truncated tetrahedra. The figure shows two views of a pentatope truncated through the mid-points of its edges. It is composed of 5 tetrahedra and 5 octahedra.

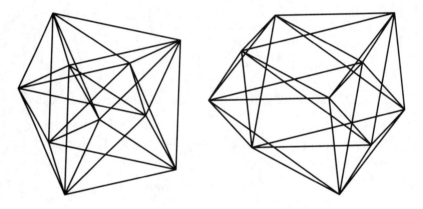

perimeter bisectors   The medians of a triangle bisect its area and so do lines parallel to one side and dividing the other side in the ratio $(\sqrt{2} + 1) : 1$. These six lines are concurrent in threes:

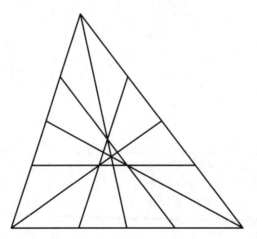

All the lines which bisect the area of a triangle, in this case an equilateral triangle, envelope three hyperbolic arcs:

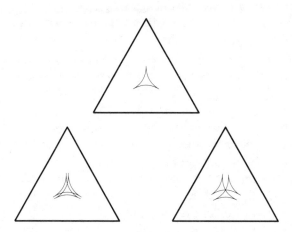

For lines which divide the triangle into unequal areas, the envelope is more complex. As the ratio changes from 1 : 1 (the top figure), it first separates as the bottom left figure and then, at a ratio of 5 : 4 three of the arcs concur. Thereafter an increasingly large central region appears, which eventually becomes the triangle itself.

REFERENCES: J. A. DUNN and J. E. PETTY, 'Halving a triangle', *Mathematical Gazette*, No. 396, 1972; DEREK BALL, 'Halving envelopes', *Mathematical Gazette*, No. 429, 1980.

Peaucellier's cell   Peaucellier was a French army officer, who was the first person to solve the problem of how to draw a straight line without the use of a ruler, by means of a linkage.

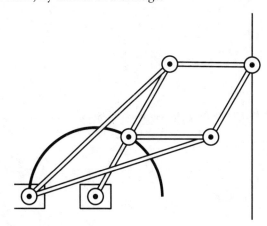

The peg at the vertex of the large vee is a fixed point. The general Peaucellier linkage just consists of the vee and the rhombus. If one vertex of the rhombus traces a circle (which passes through the fixed point) then the opposite vertex traces its inverse, which is a straight line. This is achieved by adding an extra rod and fixing its end at the centre of the circle as in the figure.

If the same vertex of the rhombus traces out, not a circle, but some other curve, then the opposite vertex will still trace out its inverse. Peaucellier's cell can therefore be used to invert any curve.

Peaucellier's invention, first published in 1867, was motivated by the common need in mechanics to transform circular motion into rectilinear motion. James Watt had previously found an approximate solution, Watt's parallel motion. A.B. Kempe, author of 'How to draw a straight line: A lecture on linkages' (1877) and inventor of several linkages himself, described how Peaucellier's cell was adapted by the chief engineer, Mr Prim, for use in the air engines which ventilated the new Houses of Parliament, and recommended a visit to his readers.

Philo's line   Given two lines forming an angle, and a fixed point X within the angle, the shortest line segment AB through the point X is called Philo's line, after Philo of Byzantium, an expert on mechanics and hydraulics who came across the idea while trying to duplicate the cube. If OY is the perpendicular from O to AB, then AX = YB.

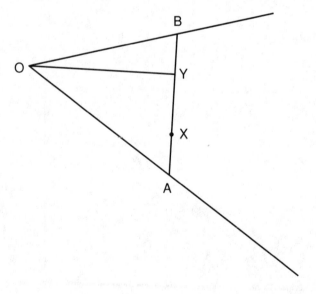

A much simpler problem is to find the line through X which encloses, with the two original lines, the smallest area. The answer is simply to construct the line which has X as its mid-point.

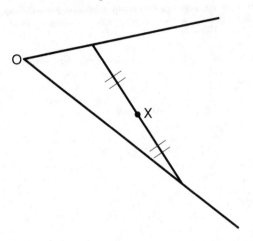

Pick's theorem In 1899 G. Pick discovered a simple method of finding the area of a polygon whose vertices lie on the points of a plane square grid. If N is the number of points of the lattice inside the polygon, and B is the number of lattice points on the boundary, including the vertices, then

$$\text{area} = N + \tfrac{1}{2}B - 1$$

In this example $N = 4$ and $B = 6$, so the area is $4 + \tfrac{1}{2} \times 6 - 1 = 6$.

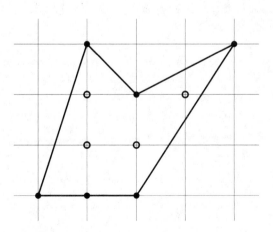

Pick's theorem is equivalent for a plane map to Euler's relationship for a polyhedron: vertices + faces = edges + 2.

pivot theorem Take any three points, A', B' and C' on the sides of a triangle ABC. Then the circles AB'C', BC'A' and CA'B' have a common point.

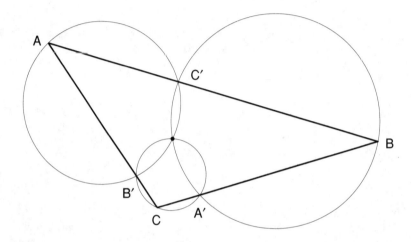

This leads to a porism. Take any three circles through a common point. Then there are an infinite number of triangle with vertices one on each of the circles, and whose sides pass through the other intersections of the circles.

There is also a three-dimensional analogue of the pivot theorem. Take six points on the edges of a tetrahedron. The four spheres each passing through a vertex and the three added points on the edges through that vertex have a common point.

plaited polyhedra A. R. Pargeter, inspired by *Plaited Crystal Models* (1888) by John Gorham, developed methods to plait many more polyhedral models, including the Platonics. 'Plaiting' refers literally to the process used to plait long hair.

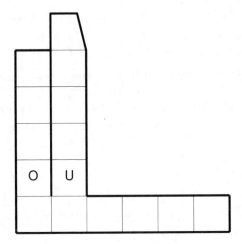

Heavy lines indicate cuts. The first pattern plaits into a cube, if the outer surface of the cube is held upwards and the initial move is to place square O over square U. The final square which tucks in to secure the plait is slightly tapered.

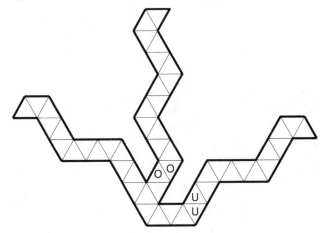

The second pattern forms an icosahedron, under the same conditions. There are two ends to be tucked in, each formed from a pair of triangles. REFERENCE: A.R. PARGETER, 'Plaited Polyhedra', *Mathematical Gazette*, No. 344, 1959.

Plateau's problem Lagrange proposed the following problem: To determine the minimal surface with a given closed boundary, and with no singularities on the surface within the boundary.

This is now called Plateau's problem, because the Belgian physicist J.A.F. Plateau solved some cases experimentally. It is in most cases an extremely difficult problem in the calculus of variations; Jesse Douglas won the first Fields Medal, the mathematicians' Nobel Prize, in 1931, for proving that in general a solution does exist. However, many approximate solutions can be obtained in practice by the use of soap films.

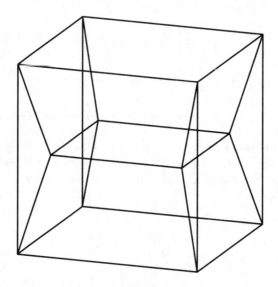

If a cubical wire framework is dipped into a soap solution, 13 surfaces will be formed. Each surface is almost plane, and they meet in threes at angles of 120°. (This is a property of all soap bubbles. It is also true that when four edges meet at a corner, they meet at equal angles, of approximately 109° 28′ 16″, the tetrahedral angle.)

H.A. Schwarz solved Plateau's problem for a skew quadrilateral in 1865, and illustrated his solution with three models, made of thin wire and covered with a film of gelatine.

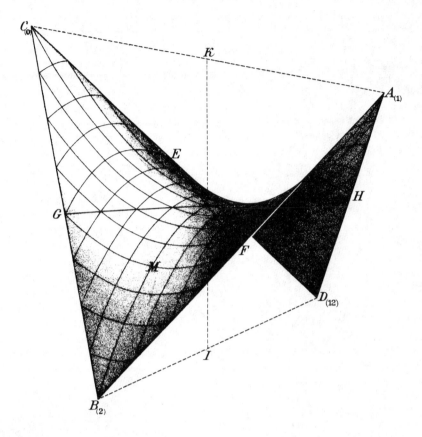

Platonic solids A polyhedron is *regular* if it has as its faces just one type of regular polygon, and all its vertices are congruent. There are only five: the cube, regular tetrahedron, regular octahedron, regular dodecahedron and regular icosahedron.

The regular polyhedra are called 'Platonic' by tradition, though the last book of Euclid's *Elements* states, 'In this book, the thirteenth, are constructed the five figures called Platonic, which however do not belong to Plato. Three of these five figures, the cube, pyramid and dodecahedron, belong to the Pythagoreans, while the octahedron and icosahedron belong to Theætetus.'

That the dodecahedron was discovered early is not surprising, since iron pyrites crystals often occur as almost regular dodecahedra and fine examples are found in southern Italy. Artificial dodecahedra have been found in Italy dating from before 500 BC.

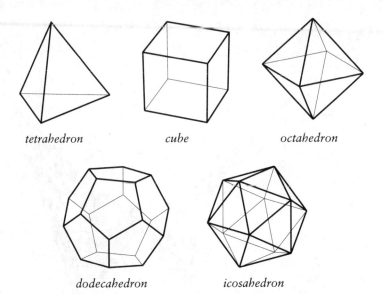

tetrahedron *cube* *octahedron*

dodecahedron *icosahedron*

The Platonic solids all satisfy Euler's relationship, that

$$\text{vertices} + \text{faces} = \text{edges} + 2$$

as can be seen from the table.

	Vertices	*Edges*	*Faces*
TETRAHEDRON	4	6	4
CUBE	8	12	6
OCTAHEDRON	6	12	8
DODECAHEDRON	20	30	12
ICOSAHEDRON	12	30	20

Poincaré's model of hyperbolic geometry Poincaré discovered that the inside of a fixed circle provides a model for *hyperbolic geometry*. In this model a line in hyperbolic geometry is an arc of a circle, within the fixed circle, whose ends are perpendicular to the fixed circle. Diameters of the fixed circle are included.

Two such arcs which do not meet correspond to parallel lines. If they meet on the fixed circle they are a pair of limit rays. Poincaré's model preserves angles, so angles can be measured directly from the figure. Arcs meeting inside orthogonally correspond to perpendicular lines.

Lengths are not preserved, however. As you get nearer to the boundary, equal lengths are represented by shorter and shorter arcs of circles, making the boundary, as it were, an infinite distance from the centre.

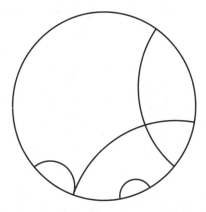

This figure shows the hyperbolic plane dissected into an infinite number of congruent triangles. In other words, all these triangles, including the infinite number of ever smaller triangles towards the edge of the disc, are the same size as well as the same shape.

pole *and* **polar**   If two tangents to a conic at A and B meet at P, then P is called the pole of the line AB, with respect to the conic, and AB is the polar of the point P.

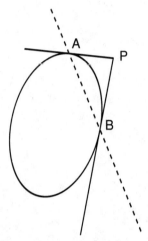

The idea of the pole and polar is a generalization of the idea of the point on the curve and the tangent at that point. Any point has a polar with respect to a general algebraic curve, and every line has a pole. If the point lies on the curve, then the pole is the tangent at that point.

Here are three of the many properties of poles and polars. A line through P meets the conic at X and Y and its polar line, AB, at Q. Then X and Y, and P and Q are *harmonic conjugates*: that is, X and Y divide the segment PQ internally and externally in the same ratio. P and Q also divide the segment XY in the same ratio.

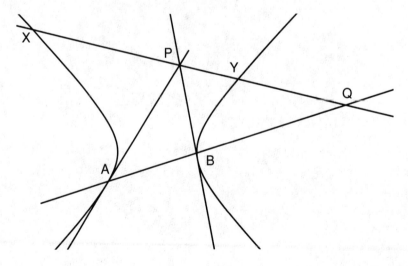

Two lines through the pole P meet the conic at Q and R, and at S and T. Then the lines QT and SR meet on the polar, and so do the lines QS and RT.

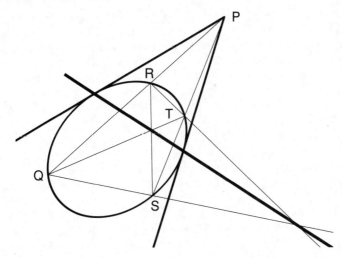

If the pole of X passes through Y, then the pole of Y passes through X. This provides a method of constructing the pole of a point inside the conic.

polygonal knots   Tie an ordinary knot in a strip of paper, carefully tighten it as you press it flat, and a regular pentagon appears:

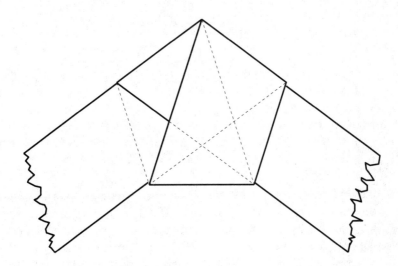

Hexagons, heptagons and larger polygons can also be folded from knots, as can be seen by thinking about the diagonals of regular polygons:

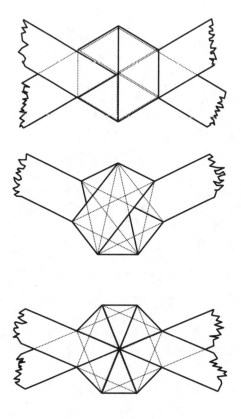

REFERENCE: H. M. CUNDY and A. P. ROLLETT, *Mathematical Models*, Oxford University Press, Oxford, 1961.

Poncelet's porism   Given two conics, for example two circles, as in the next figure, then if it is possible to draw a triangle inscribed in one to touch the other, there are an infinite number of such triangles.

For two circles, the condition for this to occur is that $R^2 - 2Rr = d^2$, where R and r are the radii of the large and small circles, respectively, and d is the distance between their centres. This is simply the relationship between the radii of the circumcircle and incircle of any triangle and the distance between their centres. (Poncelet's porism implies that if two circles are the circumcircle and incircle of a triangle, there are an infinite number of other triangles of which they are also the circumcircle and incircle.)

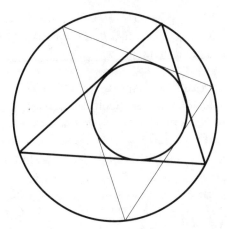

Similarly, if a quadrilateral (or *n*-gon) can be inscribed in one conic and circumscribed about the other, then there are an infinite number of such quadrilaterals (or *n*-gons).

Pons asinorum   In an isosceles triangle the base angles are equal, and if the equal sides are extended, the angles under the base are also equal. Here is Euclid's own figure, with the lines he used in his proof added.

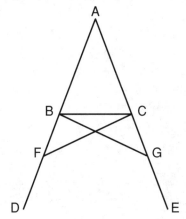

This is the fifth proposition in the first book of Euclid's *Elements*. Thales is supposed to have proved it first. Pappus proved it by, in effect, picking the triangle up, turning it over, and laying it down on itself, a method which was rediscovered a few years ago by a theorem-proving computer program.

The name means 'bridge of fools', probably referring to its resemblance to a truss bridge, and also to the fact that the weak and feeble could not get past this point in their mathematical studies.

pretzel transformations   Imagine that the object at the top left of the figure below is made out of an extremely elastic material, so that it can be stretched or squashed as much as we wish but may not be torn or cut. It might seem impossible to transform the first object into the last, without breaking one loop or pulling it through the other, but this is not so – as the sequence shows.

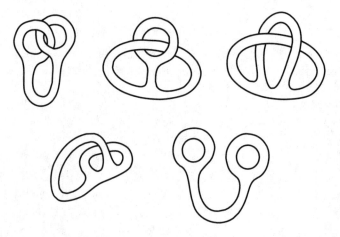

Having seen how the magic works once, it may be easier to accept this second transformation in which one of the left-hand loops just happens to fall off the large loop and hang free.

Prince Rupert's cube   How large a hole of square cross-section can be cut through a given cube? This problem is named after Prince Rupert, nephew of Charles I of England, and commander of the Royalist forces in the English Civil War. He was elected to the newly formed Royal Society; he invented an alloy, still called Prince's metal, and investigated the properties of rapidly cooled glass drops. He ended his years as Governor of Windsor Castle, where he had his own forge and laboratory.

The problem is the same as asking for the largest cube which can be passed through a given cube. Curiously, the solution is actually larger than the original cube, though only by a very small amount. If the edge of the original cube is 1, then the edge of the largest penetrating cube is $\frac{3}{4}\sqrt{2}$, or approximately 1·060 660.

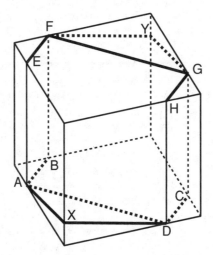

In the figure, the hole cuts the top face of the cube along the lines EFGH, the bottom face along ABCD, and the other two vertical edges at X and Y, as indicated by the heavy lines.

projective plane   In projective geometry, a straight line contains a single 'point at infinity', where both its ends meet. In other words, the line is thought of as a closed curve, which only appears to disappear in opposite directions on the plane drawing because the drawing is finite. Moreover, all lines parallel to each other share the same point at infinity.

One consequence of this idea is that a straight line extending to infinity which would divide the plane of Euclidean geometry into two parts does not separate the projective plane, which remains in one piece.

The projective plane can be represented by a region in which opposite boundary points are identified, for example by this square, in which the points A and A′ are the same point, as are B and B′, C and C′, D and D′, and so on.

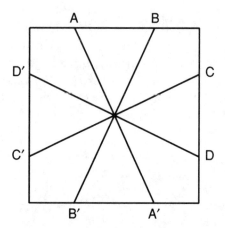

The figure below shows that a map on the projective plane may require as many as six colours, if any two regions with a common boundary are to be of different colours. Each of the six regions is adjacent to the other five.

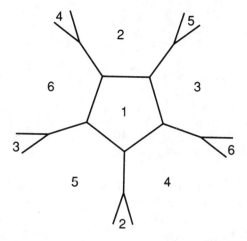

A model of the projective plane as a closed surface can be constructed by identifying both sides of the square above. However, both pairs of opposite edges will have to be given a half-twist as they are joined (in contrast to the *Möbius strip* for which only one pair of opposite edges is twisted).

First the square is stretched into a half sphere. Opposite ends of diameters are then joined as shown in the second figure to correspond to joining the points A to A′ and so on from the square. It is impossible to do this in a space of three dimensions without the resulting surface intersecting itself. The result is a surface which looks much like a sphere in its lower half, with a 'cross-cap' on top:

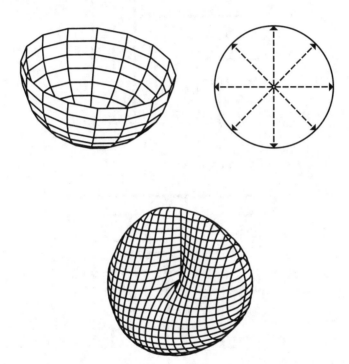

This is a one-sided surface, on which Euler's relationship becomes

$$\text{vertices} + \text{faces} = \text{edges} + 1$$

This can be checked against the map above. There are 6 regions, or faces, 10 vertices and 15 edges, which fits the formula.

There is an algebraic surface with the same form. Its equation is

$$(px^2 + qy^2)(x^2 + y^2 + z^2) = 2z(x^2 + y^2)$$

where p and q are suitable constants.

proof by looking Many simple arithmetical facts can be proved 'at sight', by examining a suitable figure.

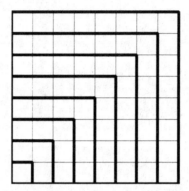

The sum of the first n odd numbers is n^2. Each odd number is represented by an L-shaped strip of unit squares.

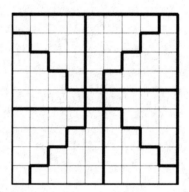

If T_n is the nth triangular number (the nth number in the sequence 1, 3, 6, 10, 15, 21, 28, 36, ...) then $8T_n + 1 = (2n + 1)^2$. Each triangular number is represented by a staircase, since the nth triangular number is equal to 1 $+ 2 + 3 + 4 + ... + n$.

The next figure shows two 2×2 squares, three 3×3 squares, and so on, thereby neatly illustrating on the flat plane that the sum of the cubes of the integers is given by

$$1^3 + 2^3 + 3^3 + 4^3 + 5^3 + ... = (1 + 2 + 3 + 4 + 5 + ...)^2$$

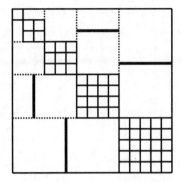

pseudosphere The Italian geometer Eugenio Beltrami realized in 1868 that the surface of the pseudosphere provided a model of a part only of hyperbolic non-Euclidean space. (There is no surface without exceptional points that is a model for the whole of hyperbolic space.)

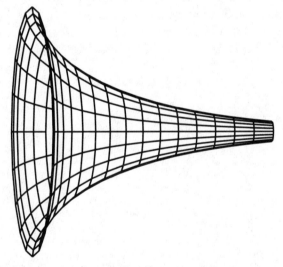

The pseudosphere is constructed by revolving a *tractrix* about its axis. A hyperbolic straight line corresponds to a geodesic on the pseudosphere. The distance between points is the distance measured along the geodesic.

Congruent figures, whose angles and lengths are the same, can be superimposed by sliding them over the surface of the pseudosphere. From our view point, the figure will appear bent, but not otherwise deformed.

The pseudosphere is a surface of constant negative curvature. Any other such surface will do as a model of hyperbolic geometry. The next figure shows on the left a surface of constant negative curvature, a surface which is cut off top and bottom by circles. In the middle is the pseudosphere, which can be continued upwards to infinity, and on the right a surface cut off below by a circular edge and with a vertex at the top.

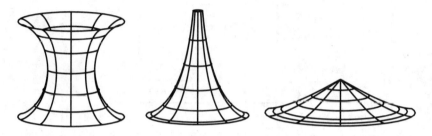

There is a simple but remarkable relationship between the trigonometry of the surfaces of constant negative curvature and those of the sphere, which has constant positive curvature. In the formulae of spherical trigonometry, leave the angles unchanged, and multiply the lengths of the sides by i, the square root of minus one.

Ptolemy's theorem   If ABCD is a cyclic quadrilateral, then

$$AB.CD + BC.DA = AC.BD$$

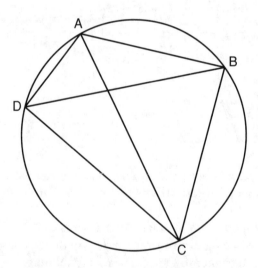

A special case (which is useful for finding *Fermat points* and Steiner trees) occurs when three of the vertices form an equilateral triangle, EFG. Then, if P is any point on the arc EF, PG = PE + PF.

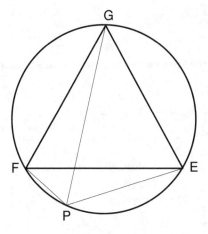

pursuit curves   Imagine four dogs chasing each others' tails, starting from the corners of a square. The path of each dog will be an *equiangular spiral*. This would still be true if the appropriate number of dogs started one from each corner of any regular polygon.

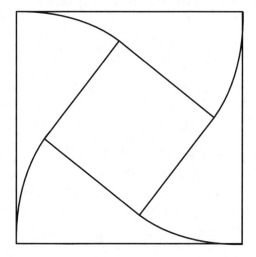

Drawing the sides of each polygon produces figures which were very popular in the more mathematical 'op-art' paintings of the 1960s because they have a strong illusion of depth (as the next figure shows).

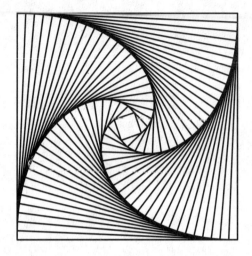

Consider a target point T which moves at constant speed along a straight line, and a moving point P which at all times moves directly towards T. If P starts *anywhere* on the outermost ellipse, and T starts from a focus of the outer ellipse, then P always captures T at the same point, the centre of the ellipse.

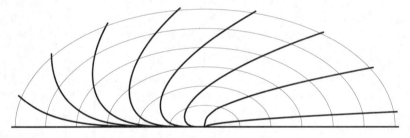

The concentric ellipses, whose shape depends on the relative velocities of T and P, are *isochrones*, and the curves of pursuit are their *isoclinal trajectories*.

Pythagoras' theorem

> In a right-angled triangle, the square on the hypotenuse is equal to the sum of the squares on the other two sides.

The most celebrated of all geometrical theorems, and the only one to feature in a popular joke, whose punch line concludes that 'the squaw on the hippopotamus is equal to the sum of the squaws on the other two hides'.

It is Proposition 47 of Book I of Euclid's *Elements*, but Euclid's proof is by no means the simplest or the easiest to follow. The theorem was once called the Theorem of the Bride and this figure is sometimes called the Bride's Chair:

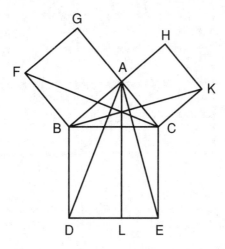

Euclid proves that triangles ABD and FBC are identical, and so are the pair KCB and ACE. Jumping ahead, he proves that the rectangle with diagonal BL is equal to the square BAGF, and similarly the rectangle with diagonal CL equals the square CAHK.

Euclid's figure has other features, which he doesn't need to use. For example, AE and BK are perpendicular, as are CF and AD, and – as Heron proved – AL, CF and BK concur.

Pythagoras' theorem appears in China at an early date. The figure on the next page is from the *Chou Pei Suan Ching* (The Arithmetic Classic of the Gnomon and the Circular Paths of Heaven), which is dated about 500–200 BC.

Far more proofs have been offered of Pythagoras' theorem than of any other proposition in mathematics. In 1940 Elisha Scott Loomis published his *The Pythagorean Proposition*, a labour of love which contained 367 proofs, including a proof by James Garfield, twentieth President of the United States, and many proofs sent in by correspondents, including several by teenagers. They were classified under four main headings and more than thirty sub-headings, but his compendium is not complete.

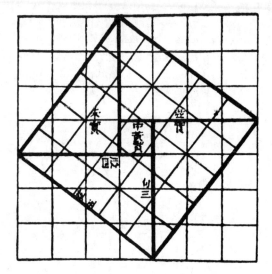

This proof is one of the simplest. In the figure below ABX + ACX = ABC, these three triangles being similar, and constructed respectively on AB, AC and BC as bases. But the areas of these triangles are in constant proportion to the areas of squares on the same bases, so the theorem follows.

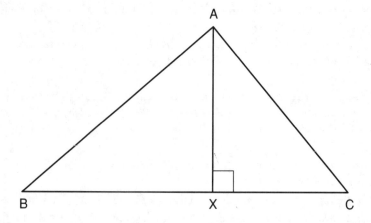

The two square tessellation (on page 260) provides another proof, by dissection. In fact, it provides an infinite number of dissections (and therefore an infinite number of proofs!) because the angled square can be placed anywhere provided it is in this orientation. The first figure shows Perigal's dissection of 1873.

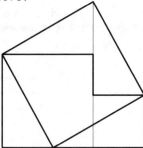

The second shows Henry Dudeney's construction of 1917.

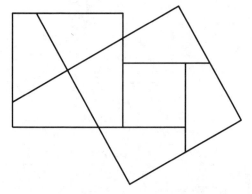

Paulus Gerdes recently suggested a very ingenious way in which the theorem might have been spotted from the same decorative motif:

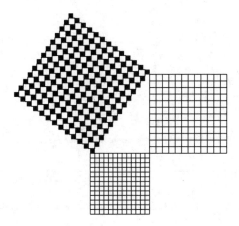

This beautiful proof is due to Leonardo da Vinci. Add a copy of the original triangle at the bottom. The figure now consists of four identical quadrilaterals. To prove that they are equal in area, imagine BA being rotated clockwise about B until BA lies along BX and the quadrilateral BAUY has become the quadrilateral BXVC.

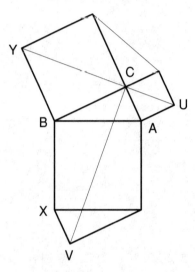

Pythagoras' theorem can be generalized in many ways. Pappus considered a scalene triangle and a line XAYZ such that XA = YZ. He constructed the three parallelograms in the figure (their angles can vary but their heights and bases are fixed), and concluded that the larger is the sum of the two smaller.

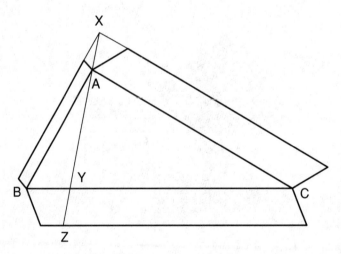

Another generalization is named after De Gua de Malves, who described it in 1783, although it was known to Descartes. Make a tetrahedron by cutting the corner off a rectangular box, so that the angles at one corner are all right angles. Then the square of the area of face ABC is equal to the sum of the squares of the other three faces.

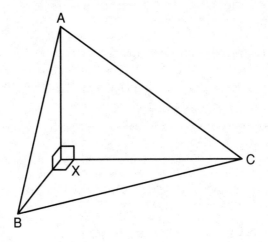

Q

quadrilateral tessellation   Any quadrilateral which does not cross itself will tessellate, even if it is re-entrant. The tessellation is related in a simple manner to the tessellation of parallelograms constructed by joining half the pairs of opposite vertices of the quadrilaterals.

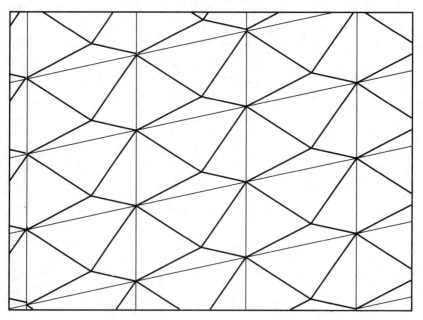

R

rectangular hyperbola   A hyperbola whose asymptotes are perpendicular is called *rectangular*.

If three vertices of a triangle lie on a rectangular hyperbola, then the orthocentre of the triangle also lies on the same curve. To put this another way, if four points are orthocentric, then there is a family of rectangular hyperbolas through the four points. The locus of the centres of these rectangular hyperbolas is the nine-point circle of the triangle.

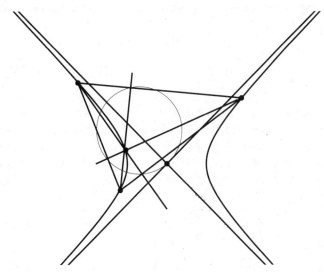

If four points are not orthocentric, then there is a unique rectangular hyperbola through them, and its centre is the point where all four nine-point circles of the triangles formed by taking three points at a time meet. If the centre of a rectangular hyperbola is taken as the centre of inversion, then the inverse curve is a lemniscate.

regular four-dimensional polytopes   These are the analogues in four dimensions of the regular polyhedra in three dimensions. There are sixteen in all, six convex and ten stellated.

The table gives the numbers of vertices, edges, two-dimensional faces and three-dimensional cells of which they are composed.

	Vertices	Edges	Faces	Cells
PENTATOPE	5	10	10	5
16-CELL	8	24	32	16
HYPERCUBE	16	32	24	8
24-CELL	24	96	96	24
600-CELL	120	720	1200	600
120-CELL	600	1200	720	120

The pentatope and the 24-cell are each their own duals. The 16-cell is the dual of the hypercube, and the 600-cell and 120-cell are duals of each other.

For all the regular polytopes an analogue of Euler's relationship holds:

$$\text{vertices} + \text{faces} = \text{edges} + \text{cells}$$

In five or more dimensions there are just three regular convex polytopes.

regular heptagon It is not possible to construct a regular seven-sided polygon using ruler and compasses only. However, it is possible to construct an angle of $\pi / 7$ by using 7 toothpicks, after which the regular heptagon is easily constructed. The toothpicks must be arranged so that A, X, Y and B lie in a straight line, and similarly on the right-hand side. The angle at A will be $\pi / 7$.

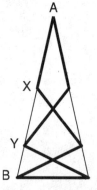

REFERENCE: C. JOHNSON, 'A construction for a regular heptagon', *Mathematical Gazette*, No. 407, 1975.

regular hexagons and stars The regular polygons which tessellate individually can be transformed into tessellations of polygons and stars by moving them slightly apart, and then dividing the space between the tiles.

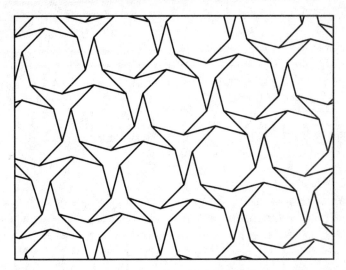

This can also be interpreted as a *hinged tessellation*. Each side of a star which is not a side of a hexagon is a strap, hinged at both ends, which holds two hexagons together. As the hexagons separate further, the star becomes fatter, then momentarily appears as a large equilateral triangle, and finally as a hexagon, identical to the original hexagons.

regular pentagon Euclid showed how to construct a regular pentagon, without which knowledge it would not be possible to construct a regular dodecahedron, as described in the last book of his *Elements*.

Many approximate constructions have been described, by Leonardo da Vinci and Dürer among many others, for the use of architects or designers. This one, drawn in the figure on the next page, is simple and perfect.

Draw a circle, and two perpendicular diameters, and divide one radius in half at X. Mark off X Y equal to X A, and, with radius A Y, centre A, draw an arc cutting the circle at B and E. Then A, B and E are three vertices of a regular pentagon.

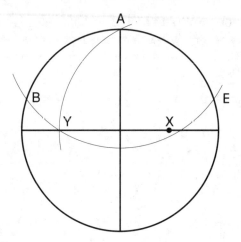

regular seventeen-gon Gauss proved, at the age of eighteen (and published in his *Disquisitiones Arithmeticæ*, 1801) that a regular *n*-gon can be constructed with ruler and compasses if *n* is a prime Fermat number, or the product of different prime Fermat numbers.

The *n*th Fermat number is $2^{2^n} + 1$, where *n* is zero or a positive integer. Since the third Fermat number is 17, it is theoretically possible to construct a regular 17-gon by ruler and compasses only. The simplest instructions are due to H.W. Richmond; they are interpreted by Rouse-Ball as follows:

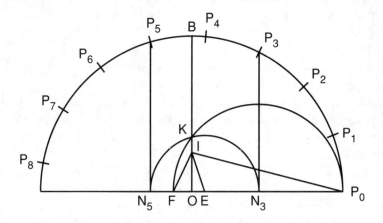

Find I on OB so that $OI = \frac{1}{4}OB$. Join IP_0 and find E and F on OP_0 so that $\angle OIE = \frac{1}{4}\angle OIP_0$, and $\angle FIE = \frac{1}{4}\pi$. Let the circle on FP_0 as diameter cut OB in K, and let the circle with centre E and radius EK cut OP_0 in N_3 (between O and P_0) and N_5.

Let lines N_3P_3 and N_5P_5, parallel to OB, cut the original circle in P_3 and P_5. Then P_0, P_3 and P_5 are the 0th, 3rd and 5th vertices of a regular 17-gon, from which the remaining vertices are easily constructed.

REFERENCE: W. W. ROUSE-BALL and H. S. M. COXETER, *Mathematical Recreations and Essays*, 12th edn, University of Toronto Press, Toronto, 1974.

regular tessellations Kepler was the first to consider the regular tessellations, recognizing them as analogues of the regular polyhedra. There are three regular tessellations using squares, regular hexagons and equilateral triangles.

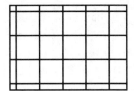 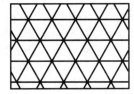 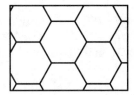

reptiles Which shapes can be dissected into identical copies of themselves? An isosceles right-angled triangle and any parallelogram with sides in the ratio $1 : \sqrt{2}$ can each be dissected into two copies of themselves.

These trapeziums each dissect into 4 copies of themselves, as does the sphinx (overpage), the only known pentagon with this property.

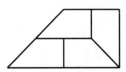 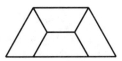

These L-shapes also dissect into 4 copies:

 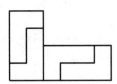

These two shapes can also be composed of smaller copies:

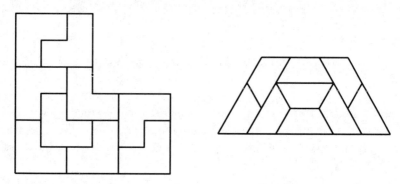

Any reptile dissection can be repeated to produce a tessellation of smaller and smaller tiles. Alternatively, the process can be reversed, and the original tiles assembled repeatedly to form a tessellation of the plane.

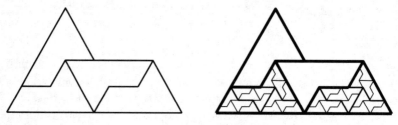

REFERENCE: C. DUDLEY LANGFORD, 'Uses of a geometric puzzle', *Mathematical Gazette*, No. 260, 1940.

Reye's configuration   Take the 8 vertices of a cube, and add the centre of the cube and the 3 'points at infinity' where the sets of parallel edges of the cube meet, as indicated by the arrows in the figure. This makes a total of 12 points.

To count 12 planes, add the 6 faces of the cube to the 6 planes passing through a pair of opposite edges. This is Reye's configuration, in which there are 12 planes and 12 points, with 6 points on every plane and 6 planes through every point.

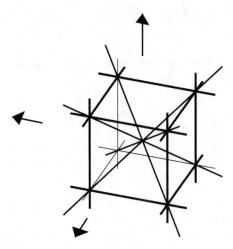

Starting with an ordinary cube is merely a convenient way of thinking about this configuration. Here is an alternative picture of Reye's configuration, without any 'points at infinity':

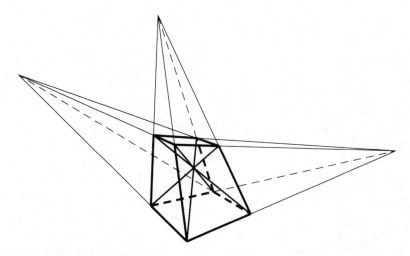

It can also be seen as a configuration of lines and points: 16 lines, the 12 edges and 4 diagonals of the cube, and the same 12 points as before. There are then 4 lines through every point and 3 points on every line.

rhombic dodecahedron Take a three-dimensional cross formed by placing six cubes on the faces of a seventh. Join the centres of the outer cubes to the vertices of the central cube. The result is a rhombic dodecahedron. Its faces are all rhombuses whose shorter diagonals are the edges of the original cube, and whose longer diagonals are the edges of a regular octahedron. It occurs in nature in crystals of garnet, among others.

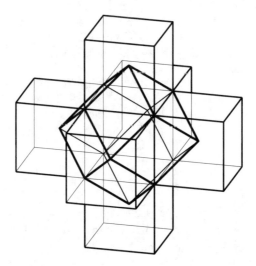

From the original method of construction, it follows that rhombic dodecahedra are space-filling. If a hinged model is made of six of the square pyramids joining the centre of a cube to one face, the model can be folded one way to make a complete cube, and the other way to make a rhombic dodecahedron, with the cubical space inscribed within it.

Faces meet either three or four to a vertex. Removing three faces that meet at a vertex and extending the six surrounding faces to form the faces of a hexagonal prism gives the form of cell found in a bees' honeycomb.

If the faces of the rhombic dodecahedron are extended until they meet each other, three stellations are formed, depending on how far the faces are extended. The first and the third stellations are shown in the figure.

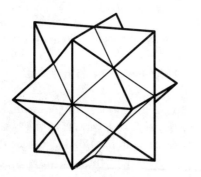

The vertices of the first stellation are the vertices of a cuboctahedron. The first stellation is also the result of interpenetrating three non-regular octahedra, each formed by compressing a regular octahedron along one of its major axes.

Careful examination of the third stellation will show that the rhombic dodecahedron is also the solid which is common to three mutually intersecting square prisms which intersect so that each pair shares a common diagonal plane. The vertices of the third stellation are also the vertices of a truncated octahedron.

rings of polyhedra The opposite edges of a regular tetrahedron are perpendicular to one another. Consequently, if made into hinges they function like a universal coupling. Provided there are enough tetrahedra to allow sufficient space inside, a ring of them can be formed which will freely rotate. For regular tetrahedra, at least 8 units are required.

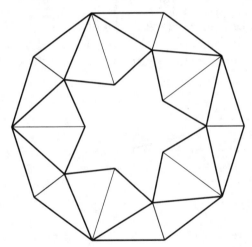

If the tetrahedra are 'longer and thinner', as few as six will rotate. A crude model can be made using pipe cleaners and straws. Take twelve straws 9 cm long and thread a pipe cleaner through each, bending the protruding ends of the pipe cleaners at right-angles to the straw and to each other. (This can be done in a right-handed or left-handed way; do six of each.) Take six more straws, each 6·5 cm long, to act as joints.

The figure shows how the straws are joined. (The short straws are not parallel to the plane of the figure.) Two long straws and two short straws make four of the six edges of a tetrahedron.

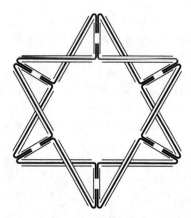

A different kind of join with extra flexibility can be achieved by using the 'wallet hinge' which appears in the well-known wallet trick, in which a currency note appears alternately above and below the straps of a wallet.

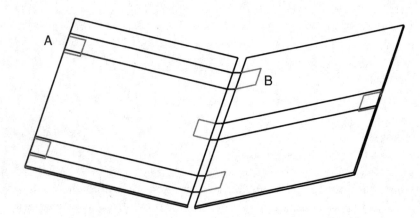

The strip from A to B is attached to the back of the left-hand card at A, and to the back of the right-hand card at B. The other strips are similarly glued.

A ring of only six cubes assembled by means of such hinges can be rotated continuously.

REFERENCE: DAVID WELLS, 'Puzzle page', *Games and Puzzles*, September 1975.

rollers   An object rolled on a circular cylinder moves smoothly without bumping because the cylinder has a constant diameter. However, a shape can have constant diameter without being a circle.

This is the *Reuleaux triangle*, an equilateral triangle with three arcs added, centred on the vertices. The constant diameter of this figure equals the side of the triangle.

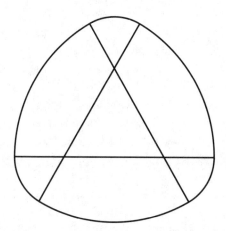

Starting from the same triangle, by drawing six arcs another roller is produced whose diameter is equal to the sum of the two radii used:

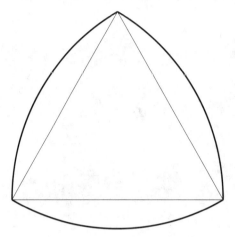

The shape on which this type of roller is based does not have to be a triangle, let alone an equilateral one: it is possible to start with any number of lines crossing. Given the four lines in the figure, place the compass point on A and draw the arc PQ. Then move the compass point to D and draw the arc QR. Then move forward to B and draw the next arc, and continue round until you return to A. The complete curve has constant diameter.

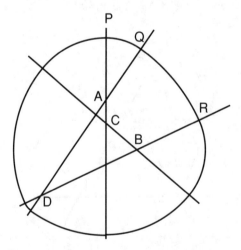

All curves of constant diameter d have the same perimeter as a circle of the same diameter, πd.

rotors   An object with rotational symmetry, such as a regular hexagon, can easily rotate inside a circle, always touching the sides. However, it is not necessary for either the rotor or the outer curve to be symmetrical.

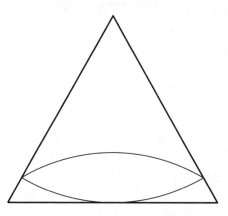

In the figure, the equilateral triangle has rotational symmetry of order three. The rotor has bilateral symmetry, each side being the arc of a circle whose centre is a vertex of the triangle and which touches the opposite side of the triangle. (The length of the rotor is equal to the height of the triangle.)

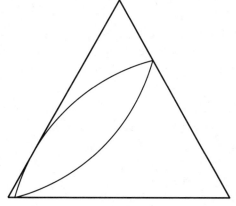

The cylinder does not have to be a circle. An equilateral triangle can rotate in this cylinder:

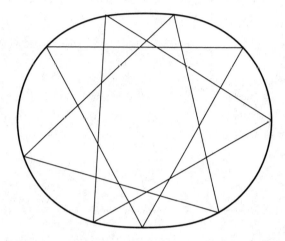

REFERENCE: H. STEINHAUS, *Mathematical Snapshots*, 3rd edn, Oxford University Press, Oxford, 1969.

S

Scherk's surface This is a minimal surface with, unusually, a very simple equation: $e^z \cos y = \cos x$. It will span four vertical parallel lines through the vertices of a horizontal square. In the figure the surface has also been cut off at the top and bottom by horizontal planes.

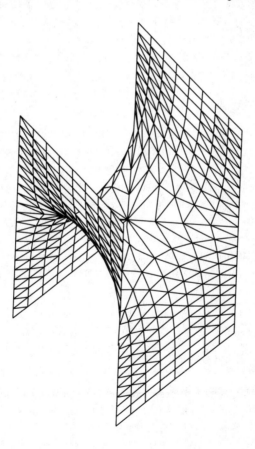

Schläfli's double six   Like *Reye's configuration*, Schläfli's double six can be most easily pictured in relation to a cube. In the next figure (showing two views of the double six) there are 30 points, with 2 lines through each point, and 12 lines, with 5 points on each line. There are 2 lines lying in each of the 6 faces of the cube, and all the lines could be coloured, say red and green, so that each red line only intersects green lines, and vice versa.

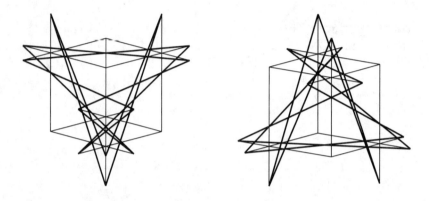

The existence of Schläfli's configuration may be expressed as the famous 'theorem of the double six'. Take a line, call it 1, and five skew lines cutting it, labelled 2′, 3′, 4′, 5′, 6′. Call the unique other line cutting 2′, 3′, 4′, 5′ line 6. Define 2, 3, 4, 5 similarly. Then the theorem states that there is a unique line, 1′ which cuts the lines 2, 3, 4, 5 and 6. The configuration that all these lines form is Schläfli's double six.

Schwarz's periodic minimal surface   Schwarz discovered two principles of minimal surfaces which allowed him to build larger surfaces from smaller units:

> If part of the boundary of a minimal surface is a straight line, then the reflection across the line, when added to the original surface, makes another minimal surface.

> If a minimal surface meets a plane at right angles, then the mirror image in the plane, when added to the original surface, also makes a minimal surface.

The boundary of Schwarz's periodic minimal surface consists of straight lines on the faces of a cube. By filling space with these cubes in the usual manner, an infinite repeating minimal surface is obtained.

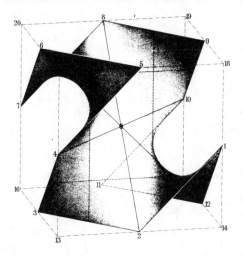

Schwarz's polyhedron   It is intuitively plausible that you can measure the area of a smooth curved surface by approximating it with many small plane triangles, and finding the limit of the area as the triangles increase in number and decrease in size.

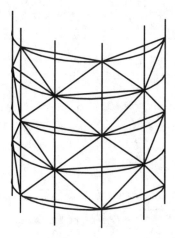

H. A. Schwarz produced this example which shows how wrong intuition can be. The basic surface is a cylinder. Divide it parallel to the cylinder's axis by $2n$ equally spaced vertical lines, and divide it at right angles to the axis by $2n^3$ equally spaced circles. Join the vertices, as in the figure, to form an accordion-pleated surface, and then let *n* tend to infinity. Instead of approximating more and more closely to the surface of the cylinder, the triangles turn against the surface, and the total surface area tends to infinity.

REFERENCE: C. STANLEY OGILVY, *Tomorrow's Math*, 2nd edn, Oxford University Press, New York, 1972.

semiregular tessellations There are 8 semiregular or *Archimedean* tessellations, all of whose tiles are regular polygons, with two or more different tiles about each vertex, and the tiling pattern around each vertex being the same. The tessellation of regular hexagons and equilateral triangles has two forms which are mirror images of one another.

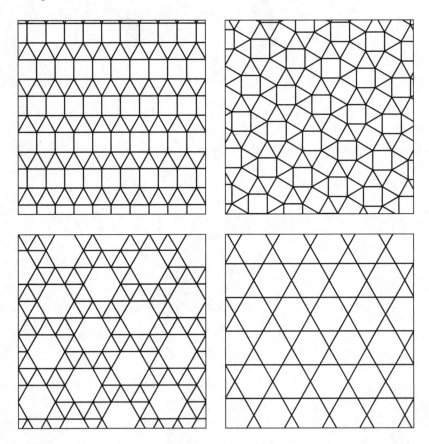

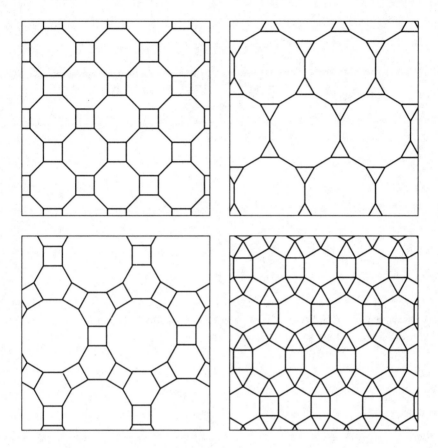

An infinite number of tessellations of regular polygons are possible if the tiling around each vertex does not have to be the same. An easy method of construction is to take one of the semiregular tessellations, move the tiles apart, and fill the gaps created with more regular polygons, just as two (at least) of the semiregular tessellations can be constructed in this manner from the regular tessellations.

seven circles theorem   Draw a circle and arrange six circles around it. They can be any size, but they must touch each other in sequence and all touch the original circle. The lines joining opposite points of contact concur.

There are many variants of the basic figure. In the illustration on the next page the original circle is at the top: five of the added circles are mutually tangent externally, but the sixth circle encloses the original circle and all the other five added circles.

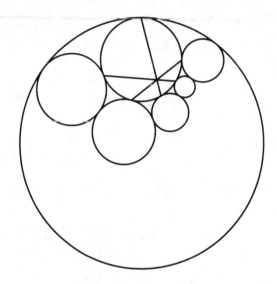

Another variant: let the radii of three of the added circles increase without limit, so that these circles become straight lines which are the sides of a triangle. The theorem is still true.

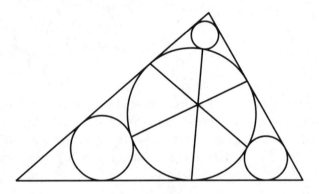

REFERENCE: C. J. A. EVELYN, G. B. MONEY-COUTTS, and J. A. TYRRELL, *The Seven Circles Theorem and Other New Theorems*, Stacey International, London, 1974.

seven colour torus A plane map can be coloured with at most four colours, so that no region is next to another region of the same colour. A map on a torus may require up to seven colours. In this map each of the regions has a common boundary with each of the six other regions, so seven colours are necessary to colour it.

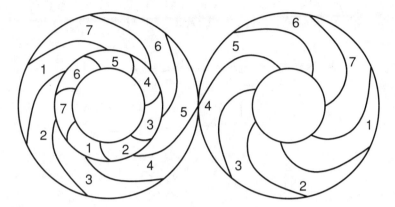

Sierpinski's square snowflake   This is Sierpinski's solution to the problem of drawing a curve which will pass through every point of a square. The figures show the first four approximations to the curve. The first two show a background of squares which are used to draw the path of the curve.

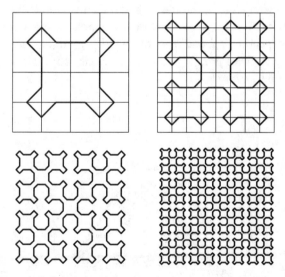

At each stage, each square is split into four quarters which are filled as in stage 1, and then joined as in stage 2 to the squares they were previously attached to. The limit of this process is a curve which passes through every point of the square.

Simson line   Take any triangle and a point P on its circumcircle. Draw perpendiculars from P to the sides of the triangle. The feet of these perpendiculars lie on a straight line, called the Simson line of that point, after Robert Simson, famous for his edition of Euclid's *Elements*.

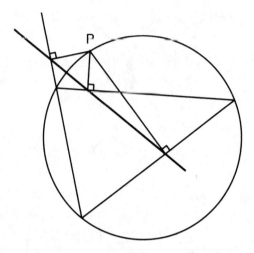

Join any point on the circumcircle to the orthocentre. The mid-point of this line lies on the nine-point circle and the Simson line of the point.

The Simson lines of two diametrically opposite points on the circumcircle are perpendicular, and meet on the nine-point circle.

Take a triangle of points on the circumcircle. Their Simson lines form another, similar, triangle.

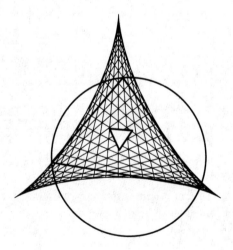

The Simson lines for all the points on the circumcircle envelope a *deltoid*. It is rather remarkable that the shape of the envelope is independent of the shape of the triangle. Each side of the triangle is tangent to the deltoid at a point whose distance from the mid-point of the side equals the chord of the nine-point circle cut off by that side. The area of the deltoid is half the area of the circumcircle of the triangle. The inscribed circle of the deltoid is the nine-point circle of the triangle.

Draw the *Morley triangle* of the starting triangle. It has the same orientation as the deltoid.

six circles theorem   Starting with a triangle, draw a circle to touch two sides, and then add another circle to touch a different pair of sides and the first circle. Continue moving round the triangle in the same way, adding more circles. The sixth circle completes the chain by touching the original circle.

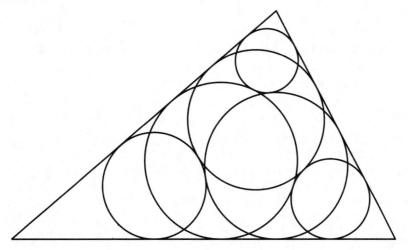

REFERENCE: C. J. A. EVELYN, G. B. MONEY-COUTTS and J. A. TYRRELL, *The Seven Circles Theorem and Other New Theorems*, Stacey International, London, 1974.

Soddy's hexlet   Sir Frederick Soddy, the chemist who discovered the laws that determine the 'chains', or series, by which radioactive elements decay into others, also discovered a remarkable chain, or necklace, of spheres.

Imagine a necklace of spheres, each sphere touching two central spheres and one which encloses the necklace.

Can such a necklace of spheres always be closed, so that the last sphere touches the first, like a *Steiner chain of circles*? Soddy showed that answer is yes, wherever the first sphere is placed, and that the necklace always contains six spheres.

Moreover, the centres of the six spheres in the necklace, and their six successive points of contact, all lie in a plane, and there are two planes which touch each of the six spheres, one on either side of the necklace.

The analogy with Steiner chains of circles is indeed very close. Soddy's figure can be obtained by inverting six identical spheres arranged around a seventh equal sphere, all sandwiched between two parallel planes.

space-filling polyhedra   Cubes can obviously fill space. Of the other regular solids, only a combination of regular octahedra and tetrahedra fill space, six octahedra and eight tetrahedra filling the space about a point in a manner that can be extended indefinitely. To see this, take four cubes forming a square. Inscribe a regular tetrahedron in each one, by joining alternate vertices, to form this ring of four tetrahedra with a bottom vertex in common.

The space in the middle forms one half of a regular octahedron; perform the same operation with eight cubes stacked to make one large cube, and the interior space forms a complete octahedron. Repeat the operation over an entire space-filling of cubes, to get the space-filling of tetrahedra and octahedra.

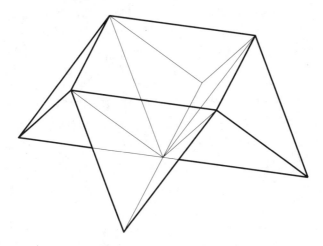

If spheres are close-packed in layers, with every alternate layer identical, and the spheres are then compressed, the result is a space-packing of rhombic dodecahedra.

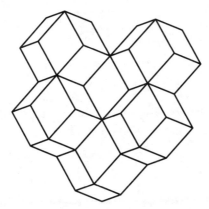

The truncated octahedron also fills space. (The volume of the truncated octahedron is half that of the cube formed by extending the truncated octahedron's square faces until they meet one another.) The next figure shows part of a row of truncated octahedra, fitted square face to square face. By adding further rows in the same horizontal plane, and then filling up the planes above and below, a network of truncated octahedra is created whose holes are identical truncated octahedra.

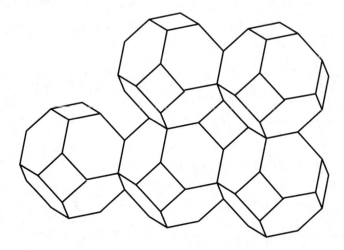

Less obvious as a space-filler is the tetrahedron with bevelled edges. In 1914 Föppl discovered a space-filling composed of tetrahedra and truncated tetrahedra. The centre of each tetrahedron is joined to its vertices, dividing it into four identical shallow triangular pyramids which are then attached to the adjacent truncated tetrahedron.

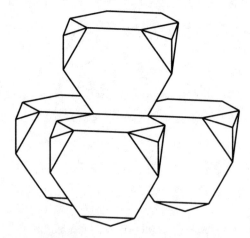

These are apparently the only space-filling solids having at least the symmetry of the regular tetrahedron. If no symmetry is required, then a convex space-filling solid can have a large number of faces, certainly as many as 38. P. Engel discovered such solids in 1980. The figure below has 18 faces and an axis of threefold rotational symmetry.

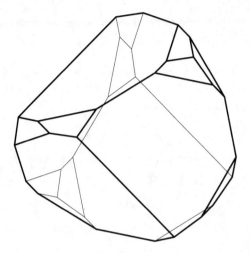

There are many other possibilities of filling space with polyhedra of two or more different types. Truncated cubes and octahedra fill space, as do the truncated tetrahedra and tetrahedra already mentioned.

This is a filling by truncated octahedra, truncated cuboctahedra and cubes, in the ratio 1 : 1 : 3.

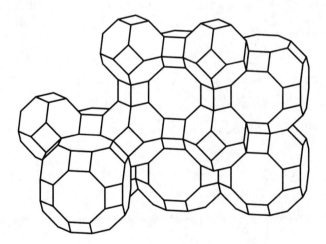

By omitting certain polyhedra from a filling a three-dimensional labyrinth may be created. Here is the space-filling of truncated cuboctahedra, truncated octahedra and cubes, with the cubes omitted:

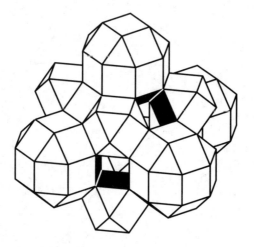

REFERENCE: P. PEARCE, *Structure in Nature as a Strategy for Design*, MIT Press, Cambridge, MA, 1978.

sphere in a cylinder   Archimedes found the volume of a sphere to be $\frac{2}{3}$ of the volume of a cylinder of the same diameter and height, and the surface area to be equal to that of the curved surface of the same cylinder.

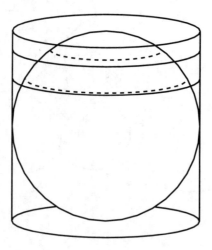

More generally, if the sphere and its containing cylinder are sliced by two planes perpendicular to the axis of the cylinder, the zones on the sphere and cylinder thus cut off have the same area.

Archimedes requested that this figure be engraved on his tombstone. Many years later, Cicero searched for Archimedes' tomb and found it, with the inscription and figure intact.

sphere packing   Place a layer of identical spheres on a flat surface, and then place another layer on top, so that there is a new sphere in every alternate dimple in the first layer. Continue to add similar layers. The result is the densest known packing of identical spheres: the spheres occupy $\pi/3\sqrt{2}$, approximately 0·7403 of space.

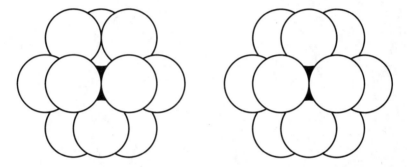

When laying the third layer, we have a choice. The spheres can be placed in one set of dimples so that they are directly over the spheres in the first layer, or so that they are above the dimples in the first layer which the second layer does *not* occupy.

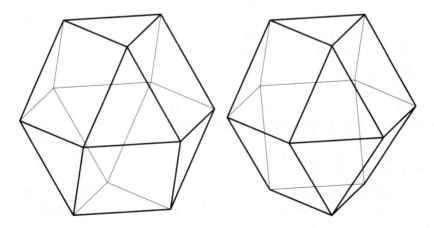

In each case a sphere in the second layer touches twelve other spheres. The first method is the most symmetrical, the centres of the twelve surrounding spheres forming a cuboctahedron. In the second case, the cuboctahedron has been sliced about an equator, and one half given a twist, to produce the second polyhedron.

Since we have a choice of two ways in which to lay every new layer, there are an infinite number of ways filling space with spheres, all having the same packing density.

spherical geometry A form of non-Euclidean geometry in which the curvature is constant and positive.

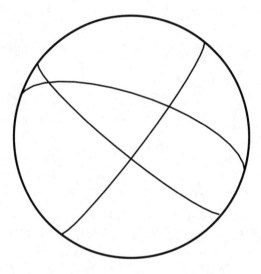

Straight lines are great circles. Any two lines meet in two points, and there are no parallel lines at all. Distances are the lengths between points as measured along the arc of a great circle, and the angle between two lines is the angle between the corresponding great circles. As in *hyperbolic geometry*, a triangle is defined by its angles, and there are no similar triangles.

The sum of the angles of a triangle is greater than two right angles, and the difference between the sum of the angles and π is a measure of the area (as in hyperbolic geometry). In the figure, if the angles of the central spherical triangle are $\pi/2$, $2\pi/5$ and $\pi/6$, then the area of the triangle is $R^2(\pi/2 + 2\pi/5 + \pi/6 - \pi) = \pi R^2/15$.

Ironically, very old results in spherical trigonometry, which go back to the Greeks, now become correct formulae in this non-Euclidean geometry!

spiral-similarity tessellation   One way to generalize the idea of a tessellation of identical tiles is to allow the tiles to be of different sizes, while remaining the same shape.

This tessellation consists of two shapes of triangle. By pairing adjacent triangles in any of three different ways, it can be seen as a tessellation of similar quadrilaterals.

Any set of corresponding points in this tessellation lies on an equilateral spiral, and all these spirals have the same pole or limiting point. The tessellation winds an infinite number of times about this limiting point, and therefore overlaps itself endlessly.

Any ordinary tessellation can be transformed into such a spiral form. Look, for example, at the six triangles, three shaded and three unshaded, which share a common vertex in this figure. They form a hexagon, and the entire tessellation can be seen as formed by such hexagons.

spirolaterals   Frank Olds derived simple rules for generating a multitude of patterns from simple instructions. Choose a starting point and a direction and follow these instructions:

```
FORWARD 1
TURN LEFT
FORWARD 2
TURN LEFT
FORWARD 3
TURN LEFT

REPEAT
```

After going four times round this loop, you return to the starting point having traced out the figure on the left. The turns are 90°, so the figure can be described as (90°: 1, 2, 3). You get the figure on the right, if you produce the spirolateral described, using the same notation, as (90°: 1, 2, 3, 4, 5, 6, 7, 8, 9) which illustrates where the name comes from.

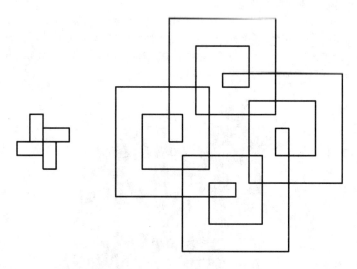

The distances moved can be 'backwards' (written as a negative number) and the turns need not be 90°. The next figures are generated from (72°: 2, 3, 4, 5) and (108°: 1, 2, 3, 4).

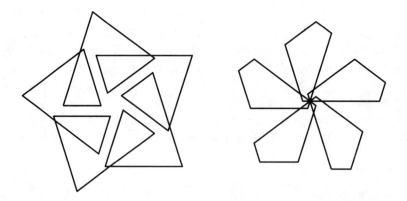

Spirolaterals include the nearest geometrical equivalent to the weird coincidences beloved of numerologists. The next figures are the same but for the different turning angle. They are (90°: 1, 3, 2, –1, –2) and (60°: 1, 3, 2, –1, –2):

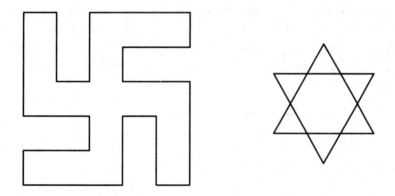

REFERENCE: MARTIN GARDNER, 'Fantastic patterns traced by programmed "worms" ', *Scientific American*, November 1973.

squared rectangles   Z. Morón was the first to dissect a rectangle into unequal squares, in 1925. The question was proposed to him by S. Ruziewicz, and appears in *The Scottish Book* as Problem 59.

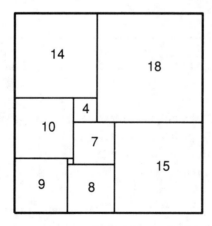

It is a curious coincidence that the first squared rectangle to be published is 32 by 33 units, and therefore close to being a squared square, which is much harder to achieve.

REFERENCE: R. D. MAULDIN (ed.), *The Scottish Book*, Birkhäuser, Boston, 1981.

squared squares The Russian mathematician Lusin once claimed that to dissect a square into unequal squares was impossible. Roland Sprague first published a dissection of a square into unequal squares, in 1939. It used 55 squares.

In 1978 A.J.W. Duijvestijn found the unique smallest simple perfect squared square, composed of 21 squares. It is perfect because all the squares are different, and simple because no subset of them form a rectangle.

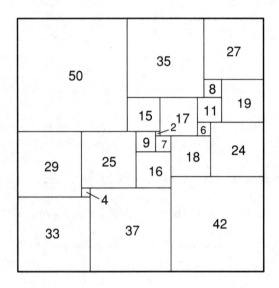

REFERENCE: R. D. MAULDIN (ed.), *The Scottish Book*, Birkhäuser, Boston, 1981.

staircase tilings Rectangular tiles from which a 'staircase' has been removed can tile the plane, by first being paired to absorb the jagged edge. This can be done in four different ways:

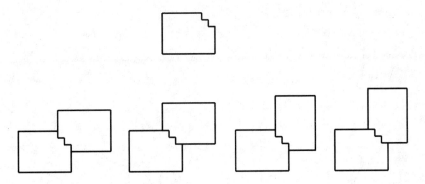

There are then many ways to tile the plane with the double pieces. This is one (the same method will tile the plane with a rectangle which has had a corner truncated at 45°):

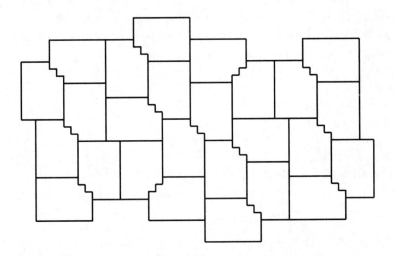

stars into polygons, stars into stars   Harry Lindgren and Greg Frederickson have been responsible for some extraordinary and beautiful dissections. The next figure shows just four of their achievements (the last one dissects into two identical stars):

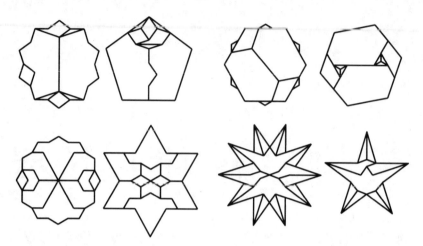

To the puzzlist, the most important feature of a dissection may be the paucity of pieces. To the mathematician, the exploitation of the natural geometry of each polygon is at least as important. These dissections possess both these virtues, plus symmetry and surprise.

REFERENCE: H. LINDGREN, *Recreational Problems in Geometric Dissections and How to Solve Them*, revised and enlarged by G. Frederickson, Dover, New York, 1972.

Steiner chains of circles   Place one circle inside another, and start a chain of circles, each touching the previous circle in the chain and the two original circles. In general, the chain will eventually overlap itself.

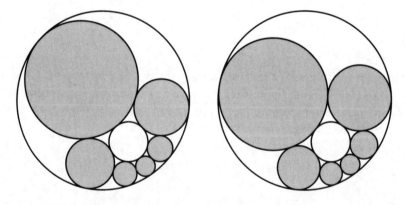

Steiner's theorem says that if the chain is closed when the last circle touches the first, then it will be closed however the first circle was drawn.

The first chain is closed. Starting with the same two circles, and placing a circle anywhere we choose, to touch both of them, the resulting chain will still close, as the second figure illustrates. The centres of all the circles of the chain lie on an ellipse, whether or not the chain closes.

It is possible that the chain will not close on the first circuit, but will close after going round several times. The theorem still holds: if one chain closes after, say, three complete turns, then any chain will close after three complete turns.

Given a general pair of circles to start with, there are thus either no solutions to the problem of fitting a closed chain of circles between them, or an infinite number of solutions. This is therefore a porism.

Steiner proved this theorem in 1826. The Japanese mathematician Ajima Chokuyen, however, had studied the same figure and come to similar conclusions in 1784.

The figure has additional properties. The tangents at the points of contact of successive circles in the chain, and the lines joining the points of contact of each circle in the chain with the outer and inner circles, all pass through one point.

Steiner networks   Steiner extended the problem of the *Fermat point* by considering four or more points and asking for the shortest route which connected all of them.

The solution for four suitably located points can be found by constructing an equilateral triangle on opposite sides of a quadrilateral.

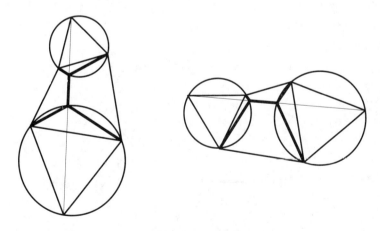

Joining the new vertices and other points of intersection, as in each of these figures, produces a network in which lines meet at 120°. (The neat point of this construction is that by choosing to draw equilateral triangles, with angles of 60°, we guarantee that the opposite angles, when they appear, will be 120° because of the property that the opposite angles of a quadrilateral inscribed in a circle sum to 180°.) Each of these solutions is a local minimum – changing it slightly will make it longer. However, it will still be true that one of these solutions is generally shorter than any other, and is the absolute minimum.

Shown below are the three possible solutions for six points at the vertices of a hexagon of unit side. The total lengths are $3\sqrt{3}$, $2\sqrt{7}$ and 5, respectively, and it is disappointing to notice that the least interesting solution is actually the shortest.

Solutions for a number of points can also be found experimentally. Arrange two plates so that they are joined by pegs representing the initial points. Dip the model in and out of a soap solution, and a soap film will form, joining the pegs together. After a few seconds it will contract into a minimum surface, under surface tension, marking a minimum route between the pegs.

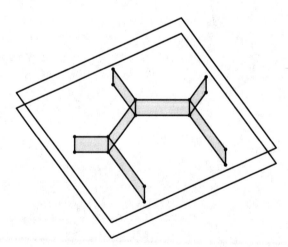

Steiner's Roman surface   Nineteenth-century mathematicians tended to be unhappy if they could not define a surface by an algebraic equation. Steiner's Roman surface is a double-sided surface with the equation

$$y^2z^2 + z^2x^2 + x^2y^2 + xyz = 0$$

The axes are double lines extending as far as the 'pinch-points' distance $\frac{1}{2}$ from the centre at the origin. It touches four circles, lying in the four planes $x \pm y \pm z = 0$. It has essentially the same form as the heptahedron.

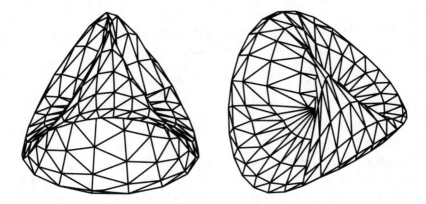

Steiner's theorem   Take two skew lines, and a line segment on each. The line segments are fixed in length, but each can slide along its own line. Joining the ends of the line segments forms a tetrahedron. The volume of this tetrahedron is constant, and does not change if the position of either segment slides along each line.

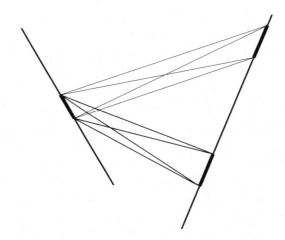

stella octangula   Imagine cutting a small octahedron from a solid block of wood by plane saw-cuts. The result will be nine pieces, the octahedron itself and eight small tetrahedra from its faces.

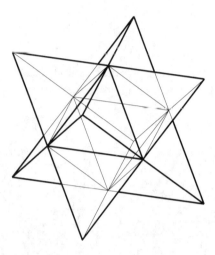

If these pieces are replaced on the faces the result is the unique stellation of the octahedron, first discovered by Kepler. It can also be thought of as the solid formed by extending the octahedron's plane faces until they meet each other again in new edges.

The *stella octangula* is also a compound of two tetrahedra, the two tetrahedra that can be inscribed in a cube by selecting alternate vertices.

T

tessellation of almost-regular polygons It is impossible to draw a tessellation composed of a mixture of regular 4-, 5-, 6-, 7- and 8-gons. The figure, from an Islamic design, shows how it is *almost* possible. Slight adjustments to the angles of the regular polygons allow tessellations such as this in which the pentagons and heptagons are not quite true.

tessellations of several squares Take a standard square grid, and allow the squares to slide apart in both directions parallel to their edges; the space between them can be made into squares of any size you choose. The result is a tessellation of two sizes of square:

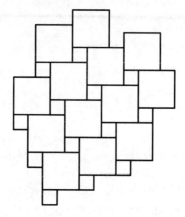

Tessellations of squares of many different sizes are also possible. This one includes three different squares:

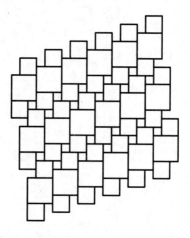

tetrahedra dissections   In two dimensions, a triangle is easily dissected into any other triangle of equal area, for example by dissecting each triangle into the same square. Indeed, any two plane polygons can be mutually dissected, if and only if they have the same area. Two tetrahedra of equal volume cannot in general be dissected into one another. In three dimensions, two polyhedra of equal volume cannot in general be mutually dissected. David Hilbert surmised that this was impossible when he presented his famous 'Twenty-three problems' during an address to the International Congress of Mathematicians in Paris in 1900. (The difficulty arises in a Euclidean space of any *odd* number of dimensions.)

This dissection of a tetrahedron into a triangular prism is one of several discovered by M. J. M. Hill and published in 1896. The hidden back edge of the base of the prism is perpendicular to the right horizontal edge and the left vertical edge. All these edges have the same length. The first cut is horizontal, one-third of the way up; the second is vertical, half-way from the left edge to the edge created by the first cut.

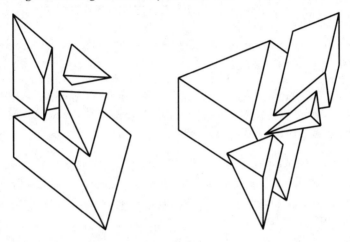

Once the prism in the second figure has been formed, it can in turn be dissected into a parallelepiped and then into a cube.

REFERENCE: V. C. BOLTYANSKII, *Hilbert's Third Problem*, Wiley, New York, 1978

tetrahedron The centre of gravity of equal weights at the vertices of a tetrahedron can be found by considering them in two pairs, and marking for each pair the point half-way between them. The two points thus found will be the mid-points of a pair of opposite sides, and the centre of gravity will lie half-way between these mid-points:

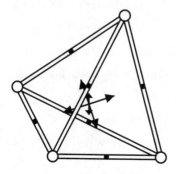

Since a pair of opposite sides can be chosen in three ways, it follows that the three lines joining mid-points of pairs of opposite sides concur. Moreover, as the second figure suggests, this becomes a theorem about the mid-points of the sides and diagonals of a quadrilateral, when the tetrahedron is projected onto a plane.

Eight spheres touch the four faces of a general tetrahedron, 1 inscribed and 7 escribed. For a regular tetrahedron 3 of the escribed spheres have their centre at infinity.

The altitudes of a general tetrahedron do not concur. They do intersect if opposite edges are perpendicular. Also, if one pair intersect then so do the other pair, and if three of the altitudes concur then all four concur. This follows from the elegant result, published by Jakob Steiner in 1827, that any line which intersects three of the altitudes of a general tetrahedron also intersects the fourth.

Thébault's theorem In 1937 Victor Thébault, a famous connoisseur of elementary and not so elementary geometry, published the result that if you construct squares on the sides of any parallelogram, their centres form another square.

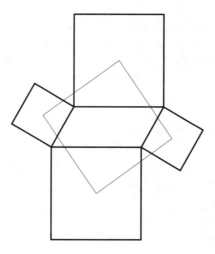

three squares into one The first treatise on dissections was written by Muhammed Abu'l-Wefa. The next figure is his dissection of three identical squares into one.

The same principle works if the half-squares on the outside are of a different size. They can also be thought of as quarters of a larger square, in which case the original dissection is of two squares, one twice the area of the other, into one square.

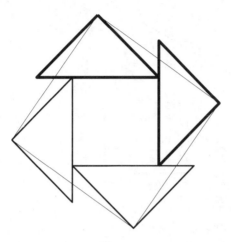

All these variants are related to the tessellation of two sizes of square. Just take such a tessellation, and mark within it one of the smaller squares and the quarters of four of the larger squares which are adjacent to it.

Here are two variations on Abu'l-Wefa's theme. In the first, a hexagon has been divided into six 120° isosceles triangles, arranged around another hexagon. This dissects any two hexagons of different sizes into one hexagon.

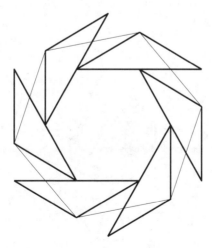

In this second variation, three congruent triangles and one similar triangle are dissected into a larger, similar triangle.

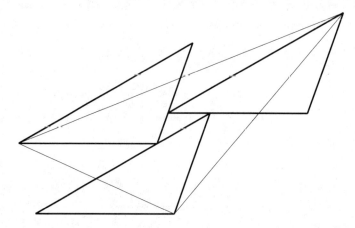

Yet another variation on the same theme is the dissection of a Greek cross into a square by joining every third vertex.

REFERENCE: DAVID WELLS, 'On gems and generalisations', *Games and Puzzles*, June 1975.

Thurston's hyperbolic paper   This model was suggested by William Thurston as a means of visualizing some of the differences between ordinary space and the space of hyperbolic geometry.

Make a surface of equilateral triangles, but fit seven equilateral triangles around each point. The surface will be floppy, and the further you extend the surface, always placing seven triangles round each vertex, the floppier it will become.

There is 'more' hyperbolic space around a point than there is Euclidean space, and if this model of hyperbolic space is flattened it becomes compressed or folded. This is just the opposite of what happens to the surface of a sphere, which stretches or tears if you try to flatten it, and which, as you get further and further away from your starting point, closes in on itself.

This is suggested by the figure below, which shows three perpendicular axes and three mutually perpendicular planes through them. These 'planes' are not completely flat even near the origin, and they crinkle and fold as they move away from it.

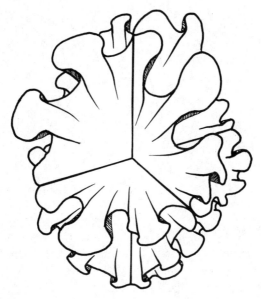

tractrix   Place an old fob-watch on a table, so that its chain just reaches the edge of the table. Pull the chain along the edge, and the path of the watch will be a tractrix, or rather one half of the tractrix. The edge of the table is the curve's asymptote. If the tractrix is rotated about its asymptote, the resulting surface is the *pseudosphere*.

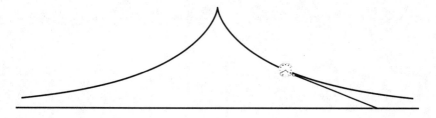

The tractrix is an involute of the catenary. Wrap a thread round one half of a catenary, ending at its vertex, and then unwrap the thread, keeping it taut. The path of the end of the thread will be the tractrix. Despite being of infinite length, the area between the curve and the axis has a finite value, $\frac{1}{2}\pi a^2$, where a is the distance of the vertex from the asymptote.

Consider the infinite set of identical circles whose centres all lie on the same straight line. The curve (apart from the line of centres, which cuts them all at right angles) is a tractrix.

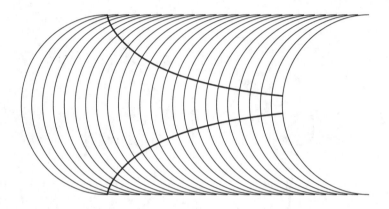

trefoil knot   This is the simplest of all proper knots, having only three crossings. It come in two distinct varieties, left-handed and right-handed. Each form can be transformed into the other by a rotation in four dimensions, as Möbius realized as long ago as 1827.

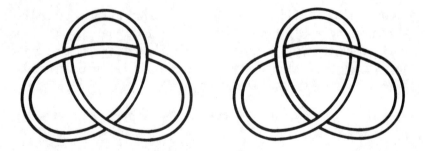

The second figure illustrates the threefold symmetry of the knot. It is equivalent to the edge of the continuous strip of paper on the right.

  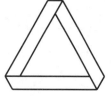

Two trefoil knots of the same handedness make a granny knot, and a pair of opposite handedness (mirror images of each other) a reef knot:

This is the shortest trefoil, and the shortest knot of any kind, which can be 'tied' with a sequence of face-to-face cubes:

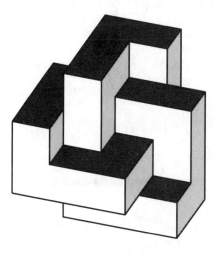

triangle reflections   Draw a triangle ABC and mark any point P. Mark the reflections X, Y, Z, of the point P in the sides of the triangle. Then the circles XYC, YZA, ZXB and ABC itself, all meet in a common point.

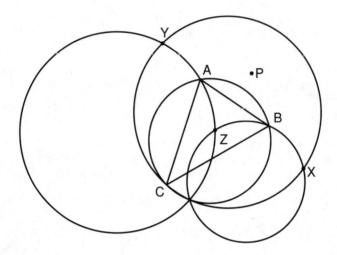

triangle tessellations   Any single triangle will tessellate, by using a pair of triangles to make a parallelogram:

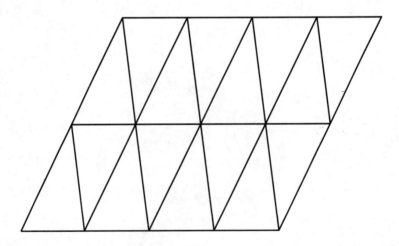

There are very many other possibilities, based on a variety of triangles, which have hardly been investigated.

Here is a tessellation of two different triangles, each of which appears in three sizes:

twenty-four cell This four-dimensional polytope is self-dual, having 24 cells and vertices, and 96 edges and faces. Each of the 24 cells is an octahedron. Each octahedron has 8 triangular faces, and each face is shared by 2 octahedra: hence the 96 faces.

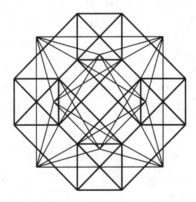

The 24-cell is a truncation of the 16-cell. It can be used to pack four-dimensional space, as can the hypercube.

A rhombic dodecahedron can be constructed from two cubes: one of them is cut into six pyramids by joining its vertices to its centre, and one pyramid is then stuck onto each face of the other cube. In an analogous manner, the 24-cell can be constructed from two hypercubes, by cutting one of the hypercubes into 8 cubic pyramids, each based on one of its 8 cubical cells.

twisted triangular prism   This figure is formed from a triangular prism, by giving it a twist and then joining the ends together. It has two faces and one edge, and is equivalent to a torus with a spiral going round it three times before returning to its starting point.

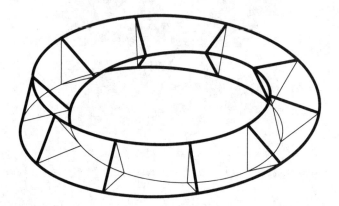

On the South Bank, in London, a sculptor has partly embedded in the ground a twisted triangular prism, with some chunks missing. Children much enjoy playing on it.

two squares tessellation   Any two sizes of square can be used to make this simple tessellation, which can also be thought of as a tessellation of large squares in which every row and column has been slid apart by the same amount to leave a pattern of identical small square holes.

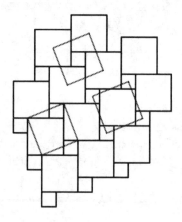

As in the Greek cross tessellation, joining any four suitably chosen corresponding points together produces a dissection of the two original squares into one larger square. Several of the best-known dissections of two squares into one correspond to obvious choices for the corresponding points, such as the centres of the large squares, or of the small squares, or the corners of the squares.

All these dissections effectively prove Pythagoras' theorem.

U

unduloids   If a pair of circular rings, parallel to each other and on the same axis, are dipped into a soap solution, the minimal surface formed when the rings are empty is the *catenoid*. If, however, the rings are replaced by solid discs, so that the pressure within the film is no longer equal to the pressure outside, the minimal surface is an unduloid.

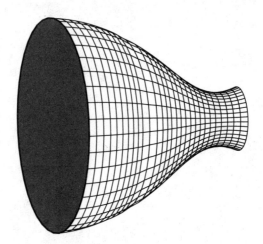

The outline of the unduloid is always part of the locus of the focus of a conic which rolls along a straight line.

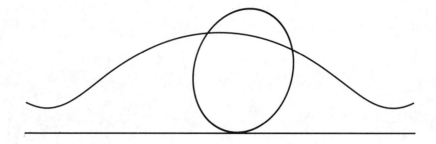

uniform polyhedra   A polyhedron is uniform if all its faces are regular (the face may be a regular star polygon) and all its vertices the same. The Platonic and the Archimedean polyhedra are the convex uniform polyhedra. Here are two others, the small cubicuboctahedron (left) and the small dodecahemicosahedron (right); which are not convex, and whose faces intersect each other.

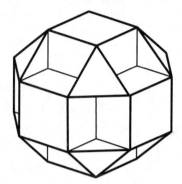 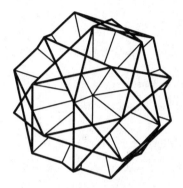

The small cubicuboctahedron is a rhombicuboctahedron with 12 square faces removed and 6 regular octagonal faces inserted. The vertices of this solid are the outer vertices, which are the vertices of the equilateral triangles. The internal points where the faces cut each other do not count as vertices.

The small dodecahemicosahedron has 12 star pentagons on the faces of a regular dodecahedron, together with 10 regular hexagons, all passing through the centre of the solid. Once again, the internal intersections of the faces are not counted as vertices.

Coxeter, Longuet-Higgins and Miller published their enumeration of the uniform polyhedra in 1954, counting 53 in addition to the Platonic, Archimedean and Kepler–Poinsot polyhedra and the prisms and antiprisms. They expressed the belief that their list was complete, but offered no proof.

unilluminable room   A room which is not convex cannot be illuminated by a single lamp at any point within it. It might be supposed, however, that if the walls were entirely covered with mirrors then every part would be illuminated wherever the lamp were placed.

Surprisingly, though, a 'hall of mirrors' of the shape shown below cannot be completely illuminated from any point within it. The shape is based on an ellipse, which has the property that a ray of light from one focus will be reflected to the other focus.

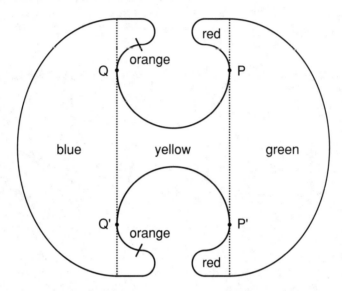

The ellipse has been cut along its major axis, and separated into green and blue halves, the foci being P, P', Q and Q'. The curves forming the other parts of the wall can be chosen with more variety, as long as they touch the major axis at the foci. A ray of light from 'behind' one of the foci will be reflected behind the other, and a ray of light crossing in front of one focus will be reflected 'in front of' the other. Thus light from one red 'corner' can illuminate only the green region and the other red corner; light from the green or yellow region will never reach the orange corner; and so on.

unistable polyhedra Is there a polyhedron which, if constructed out of a material of uniform density, would stand on one face only, and roll or fall over if placed on any other face? Richard Guy proved that there is, and that this prism of 17 sides and 19 faces is unstable.

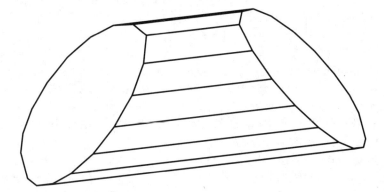

The second figure is a cross-section, showing the prism's symmetry.

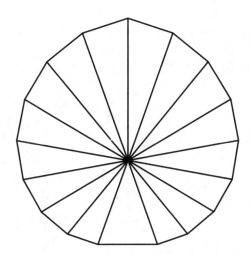

V

Verhulst process Verhulst, a pioneer in the study of population, described in 1838 a law according to which a population would not increase in size indefinitely, contrary to the pessimistic forecasts of Malthus, because obstacles to its increase would increase faster than the population itself.

Processes of the form $x_n + 1 = r\, x_n\, (1 - x_n)$ are named after him. The behaviour of the Verhulst process depends very much on the chosen value of r. If $r < 2$, then the system quickly settles down to a steady value. However, if $2 < r < 2 \cdot 5$, then the population oscillates between two values. The solution is said to have *bifurcated*.

As r increases a little beyond $2 \cdot 5$ the solution bifurcates again, then again, to oscillate between eight values, then sixteen values, and so on. These bifurcations come closer and closer together, until after an infinite number of bifurcations, when r is approximately $2 \cdot 570$, the solutions become chaotic.

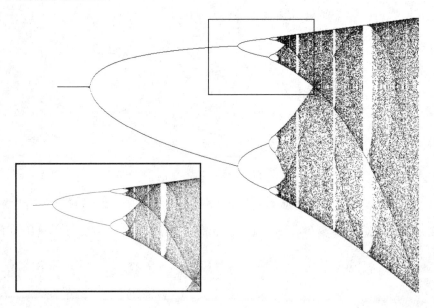

The points now start jumping around, superficially without rhyme or reason. Actually they have either no periodic behaviour, or very high periods. This chaotic region itself, however, does have structure. There are vertical bands within the region, it is criss-crossed by the previous outer boundaries which continue over it, and at about $r = 2 \cdot 83$ the whole Verhulst figure appears again, in miniature.

In order for an infinite number of bifurcations to appear in a finite interval, the distance between successive bifurcations must get less and less very quickly. They do, and the ratio of these distances tends to a limit, called Feigenbaum's number after its discoverer. It is approximately 4·669 201 660 9.

Viviani's theorem   In an equilateral triangle, the sum of the perpendiculars from any point P to the sides, is equal to the altitude of the triangle:

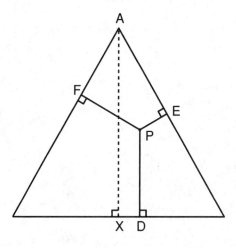

$$PD + PE + PF = AX$$

If P lies outside the triangle the relationship still holds, provided one or two of the perpendiculars (the ones that lie entirely outside the triangle) are measured as negative.

Voderberg tilings Voderberg published a description of this remarkable tile in 1936. Two tiles will completely enclose not only one other tile, but two. In the second figure, the four tiles form a decagon whose opposite edges are equal and parallel.

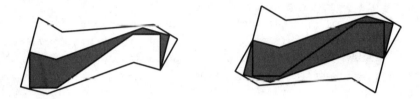

Copies can be fitted together to form a infinite horizontal strip, and duplicate strips will then join it to cover the plane. It is also the basis for this spiral tessellation with two 'centres':

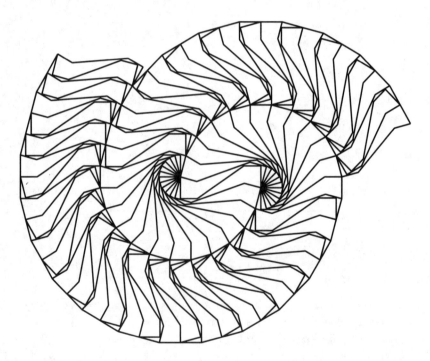

The next figure shows how one of the matching arms of the spiral is constructed. The first portion consists of 12 single tiles arranged in a half-circle around the right-hand 'centre'. The second portion consists of 3 × 12 = 36 tiles, arranged in groups of three, with 24 'facing' outwards and 12 'inwards'. The third portion, if shown completely, would consist of $5 \times 12 = 60$ tiles in groups of five, with 36 facing outwards and 24 facing inwards. The pattern continues, with each arm occupying a half-circle and formed in 12 groups of 5, 7, 9, ... tiles.

Since Voderberg's pioneering example, many tiles with similar properties have since been found. For example, Branko Grünbaum and G. C. Shephard discovered and named this *versatile* which can also be used to construct spirals with 1, 2, 3, and 6 centres, as well as many other tilings.

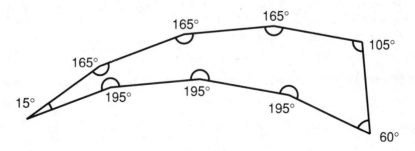

These figures illustrate one way to construct spiral tessellations based on the first figure in the set, which has rotational symmetry of order 10. This tessellation can be extended outwards for ever.

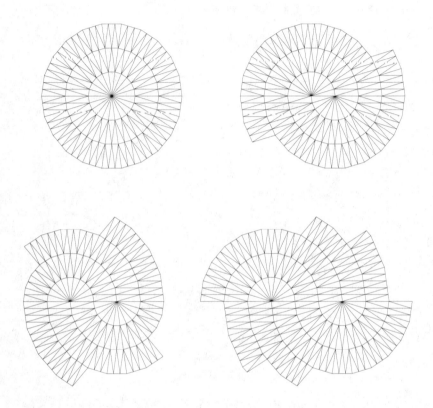

Pick any one of its diameters and slide one half of the tessellation to one side just far enough to give the second figure with one spiral arm. Next, continue to slide the two halves of the original tessellation along the diameter, the same distance again, to produce the third figure with two separate spirals. A similar movement produces the final figure with three separate arms, and the process can be continued indefinitely.

wallpaper patterns A wallpaper pattern repeats at regular intervals in two different directions. Its symmetries depend on the symmetry of the underlying network, and also on whether the repeating motif has any of the same symmetries. The network can have five forms as shown in the rows in this figure.

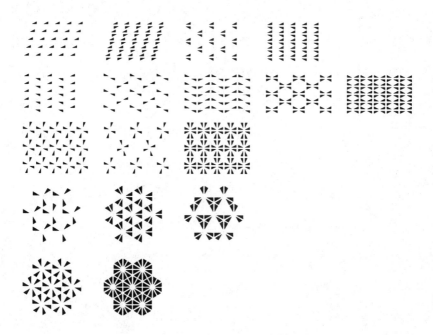

Combining the possible symmetries produces 17 types of wallpaper pattern, all of which are found among the tilings of the Alhambra palace in Spain, and elsewhere in Moorish architecture.

woven polyhedra Is it possible to wrap the surface of a polyhedron with cylindrical strips in a uniform and symmetrical manner? Yes it is, as Jean Pedersen demonstrated. Her models are similar to some traditional types of knots and Japanese decorative Temarı balls. The figure shows six strands covering the surface of an dodecahedron.

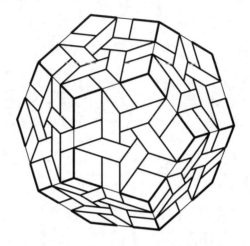

REFERENCE: JEAN PEDERSEN, 'Geometry: The unity of theory and practice', *Mathematical Intelligencer*, Vol. 5, No. 4, 1983.

Z

zonagons   If a polygon has an even number of sides, all its sides equal in length and opposite sides parallel, then it is called a zonagon. It can be dissected into rhombuses. A square, which is a rhombus already, is the only such polygon which does not have more than one such dissection. A hexagonal zonagon can be dissected into rhombuses in 2 ways, an octagonal zonagon in 4 ways, and so on.

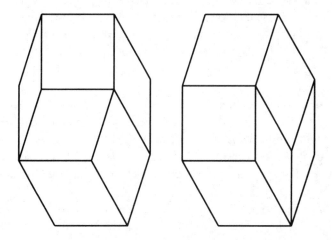

Regular even-sided polygons are zonagons, and odd-sided polygons can be transformed into zonagons and dissected into rhombuses if the midpoints of the sides are considered as extra vertices, doubling the number of sides.

These divisions into rhombuses are useful for solving dissection problems, as *dodecagon dissection* illustrates.

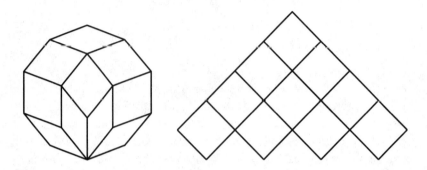

Because the elements are equal-edged these polygons can be hinged, and even turned into tessellations of squares, if the vertex at the bottom of the left hand figure, where four rhombuses meet, is first broken.

REFERENCE: JUDITA COFFMAN, 'Maths club activities', *PLUS*, No. 3, 1986.

zonahedra The zonahedra were first investigated by E. S. Fedorov, in connection with crystallography. All the edges of a zonahedron are the same length, and all the faces are rhombuses; the edges lie in n given directions only. It necessarily has $n(n-1)$ faces. (If the faces are not rhombuses, but are still equilateral and have opposite edges parallel, then the resulting figure is a parallelohedron.)

The simplest zonahedron is the rhombic prism or rhombohedron, with edges lying in three directions only. A cube is a special case of this solid. The general rhombic dodecahedron has edges in four directions only, and therefore has $4 \times 3 = 12$ faces. If the edges are in the directions of the four diagonals of a cube, it is a regular rhombic dodecahedron.

The six diameters of a regular icosahedron lead to the rhombic triacontahedron, with 30 faces, the dual of the icosidodecahedron. Removing one complete zone of rhombuses from this figure (in other words, removing all the rhombuses containing edges in one particular direction), leaves the rhombic icosahedron of 20 faces and removing a zone from this produces a rhombic dodecahedron (which is not the rhombic dodecahedron which is the dual of the cuboctahedron).

Any zonahedron can be dissected into parallelepipeds, which in turn can be dissected into cubes. Therefore any two zonahedra can be dissected one into the other provided they have the same volume.

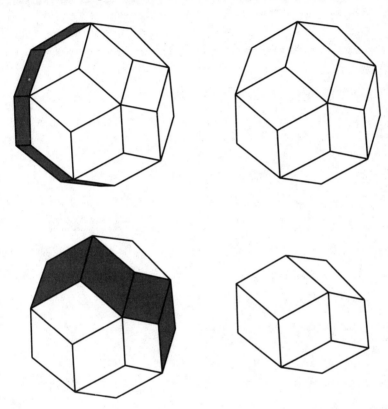

Index

Visit Penguin on the Internet
and browse at your leisure

- preview sample extracts of our forthcoming books
- read about your favourite authors
- investigate over 10,000 titles
- enter one of our literary quizzes
- win some fantastic prizes in our competitions
- e-mail us with your comments and book reviews
- instantly order any Penguin book

and masses more!

'To be recommended without reservation ... a rich and rewarding on-line experience' – Internet Magazine

www.penguin.co.uk

READ MORE IN PENGUIN

In every corner of the world, on every subject under the sun, Penguin represents quality and variety – the very best in publishing today.

For complete information about books available from Penguin – including Puffins, Penguin Classics and Arkana – and how to order them, write to us at the appropriate address below. Please note that for copyright reasons the selection of books varies from country to country.

In the United Kingdom: Please write to *Dept. EP, Penguin Books Ltd, Bath Road, Harmondsworth, West Drayton, Middlesex UB7 ODA*

In the United States: Please write to *Consumer Sales, Penguin Putnam Inc., P.O. Box 12289 Dept. B, Newark, New Jersey 07101-5289.* VISA and MasterCard holders call 1-800-788-6262 to order Penguin titles

In Canada: Please write to *Penguin Books Canada Ltd, 10 Alcorn Avenue, Suite 300, Toronto, Ontario M4V 3B2*

In Australia: Please write to *Penguin Books Australia Ltd, P.O. Box 257, Ringwood, Victoria 3134*

In New Zealand: Please write to *Penguin Books (NZ) Ltd, Private Bag 102902, North Shore Mail Centre, Auckland 10*

In India: Please write to *Penguin Books India Pvt Ltd, 11 Community Centre, Panchsheel Park, New Delhi 110017*

In the Netherlands: Please write to *Penguin Books Netherlands bv, Postbus 3507, NL-1001 AH Amsterdam*

In Germany: Please write to *Penguin Books Deutschland GmbH, Metzlerstrasse 26, 60594 Frankfurt am Main*

In Spain: Please write to *Penguin Books S. A., Bravo Murillo 19, 1° B, 28015 Madrid*

In Italy: Please write to *Penguin Italia s.r.l., Via Benedetto Croce 2, 20094 Corsico, Milano*

In France: Please write to *Penguin France, Le Carré Wilson, 62 rue Benjamin Baillaud, 31500 Toulouse*

In Japan: Please write to *Penguin Books Japan Ltd, Kaneko Building, 2-3-25 Koraku, Bunkyo-Ku, Tokyo 112*

In South Africa: Please write to *Penguin Books South Africa (Pty) Ltd, Private Bag X14, Parkview, 2122 Johannesburg*

BY THE SAME AUTHOR

The Penguin Book of Curious and Interesting Mathematics

This fascinating compendium of strange facts and anecdotes includes African river-crossing problems, monkeys and typewriters, not to mention a definitive proof that 1 and 1 makes 2. Spanning the centuries, David Wells introduces a collection of choice eccentrics who calmed their nerves with algebra or used sextants to measure the buttocks of Hottentot women. With Wells's unique gift for making mathematics lively and accessible, this book will provide both endless entertainment and a window on a weird and wonderful world.

The Penguin Book of Curious and Interesting Puzzles

Wherever there are human beings, setting and solving problems have always been among their principal passions. This collection of logical and mathematical puzzles, none requiring more than pencil, paper and a few counters, brings together examples from the earliest times up to the present day. Whether they concern Prisoner's Dilemmas, fast-breeding rabbits, liars and truthtellers or Prince Rupert's cube, one thing is sure: endless entertainment is guaranteed.

The Penguin Dictionary of Curious and Interesting Numbers

Why was the number of Hardy's taxi significant? How many grains of sand would fill the universe? What is the connection between the Golden Ratio and sunflowers? From minus one and its square root to numbers so large that they boggle the imagination, all you ever wanted to know about numbers is here. There is even a comprehensive index for those annoying occasions when you remember the name but can't recall the number. Kaprekar numbers? Ah, yes, of course . . .

You Are a Mathematician

This entertaining and informative introduction to mathematics begins with the secrets of triangles and the dazzling patterns formed by even the simplest numbers. It takes readers on 'a journey from the Greek mathematicians to quantum theory', and concludes with a challenging game. Mathematics is an invaluable scientific tool, yet mathematical thinking is very like a game, relying on cunning tactics, deep strategy and brilliant combinations – this book is an ideal guide to its potential and pleasures.